Seeding Solutions

Volume 2

Funding Organizations for the Crucible II Project

Canadian International Development Agency (CIDA), Canada
Dag Hammarskjöld Foundation (DHF), Sweden
German Federal Ministry for Economic Cooperation and Development/
German Technical Cooperation (BMZ/GTZ), Germany
International Development Research Centre (IDRC), Canada
Swedish International Development Cooperation Agency (Sida-SAREC), Sweden
Swiss Agency for Development and Cooperation (SDC), Switzerland

Partner Organizations

Action Group on Erosion, Technology and Concentration (ETC Group), Canada
(formerly Rural Advancement Foundation International (RAFI))
International Plant Genetic Resources Institute (IPGRI), Italy

Seeding Solutions

Volume 2. Options for national laws governing control over genetic resources and biological innovations

The Crucible II Group

Copublished by
the International Development Research Centre
the International Plant Genetic Resources Institute
and the Dag Hammarskjöld Foundation

INTERNATIONAL PLANT GENETIC RESOURCES INSTITUTE
Via dei Tre Denari, 472/a, 00057 Maccarese (Fiumicino), Rome, Italy
(http://www.ipgri.cgiar.org)
ISBN 91-9043-499-6

DAG HAMMARSKJÖLD FOUNDATION
Övre Slottsgatan 2, 753 10 Uppsala, Sweden (http://www.dhf.uu.se)
ISBN 91-8521-433-7

INTERNATIONAL DEVELOPMENT RESEARCH CENTRE
PO Box 8500, Ottawa, ON, Canada K1G 3H9 (http://www.idrc.ca)

Canadian Cataloguing in Publication Data

Crucible II Group

Seeding solutions. Volume 2. *Options for national laws governing control over genetic resources and biological innovations*

Copublished by the International Plant Genetic Resources Institute and the Dag Hammarskjöld Foundation.
Includes bibliographical references.
ISBN 0-88936-958-8

1. Germplasm resources, Plant.
2. Plant varieties — Protection.
3. Plants, Cultivated — Patents.
4. Biological diversity conservation.
5. Patents.

I. International Plant Genetic Resources Institute.
II. Dag Hammarskjöld Foundation.
III. International Development Research Centre (Canada).
IV. Title.
V. Title: Options for national laws governing control over genetic resources and biological innovations

K1401.C78 2001 333.95'34 C2001-980312-9

Contents

Preface

The Crucible Groups

In 1993, after the UN Conference on Environment and Development (UNCED) and before finalization of the General Agreement on Tariffs and Trade (GATT) Uruguay Round of Multilateral Trade Negotiations, a group of 28 individuals from 19 countries met, first in Rome, and then in Uppsala and Bern. Members of the group came from South and North, from the private and public sectors, and from civil society organizations. Some were scientists, while others were policy- and opinion-makers or business executives. Although these individuals, who dubbed themselves the Crucible Group, held vastly differing views on many controversial issues, they shared a concern for the conservation and enhancement of plant genetic resources and an alarm that decisions were being taken or policies adopted that could imperil the availability of these resources for world food security and agricultural development.

In an effort to clarify issues and choices for decision-makers, the Crucible Group agreed to debate the most contentious points among themselves and to prepare a non-consensus report that would simply set forth the best arguments of every side. Following many months of e-mail and face-to-face debate, the group identified ten areas where no agreement was possible but where they could offer distinct viewpoints that might help others. Most of these issues involved intellectual property related to living organisms, the role of the Consultative Group on International Agricultural Research (CGIAR) and the future structure of an international genetic resources conservation and exchange system.

To its own surprise, however, the group identified 28 recommendations they felt able to offer collectively to policy- and opinion-makers. In June 1994, *People, Plants, and Patents* was released at a seminar hosted by the International Development Research Centre (IDRC) in Ottawa. After the release of the book, many group members followed up with seminars and workshops at the Biodiversity Convention meetings in Nairobi, Nassau, Djakarta and Montréal, and at Food and Agriculture Organization of the United Nations (FAO) meetings in Rome and Leipzig. The book was translated into French and Spanish and was widely distributed.

Five years after their first full session at the Dag Hammarskjöld Foundation (DHF) in Uppsala, many of the same people found themselves together again in the same place, revisiting the same unresolved issues. Although the group that met in 1998 had not entirely planned this, they quickly agreed that there was a need to convene 'Crucible II' and to try to advance the international agenda for genetic resources. Efforts were made to broaden the group's membership, and the style of dialogue was updated to take better account of the Internet and negotiations via e-mail. By the end of this second Crucible round, in addition to the first Uppsala meeting hosted by DHF, Crucible participants had gathered at large-scale working meetings in Ottawa, hosted by IDRC; in Nairobi, hosted by the African Centre for Technology Studies (ACTS); and in Rome, hosted by the International Plant Genetic Resources Institute (IPGRI). Six years after the publication of *People, Plants, and Patents,* the Crucible II Group published *Seeding Solutions. Volume 1. Policy options for genetic resources:* People, Plants, and Patents *revisited*. One year after the release of Volume 1, we are now pleased to be releasing *Seeding Solutions. Volume 2. Options for national laws governing control over genetic resources and biological innovations.*

The Crucible II Group, now even more than in its first round, continues to be a diverse gathering of individuals who passionately and respectfully disagree on intellectual property, the rights of farmers, the mechanisms for benefit-sharing, and the appropriate structures for conservation. More than 45 individuals from 25 countries took part in one or more of the face-to-face discussions and exchanged opinions and data electronically. As with the original Crucible members, Crucible II also shares a passion for plant genetic resources and an ever-growing alarm that one of humanity's vital resources is being threatened or squandered. Beyond this, the members have 'agreed to disagree' and have worked hard and cooperatively to describe their differences without compromise.

Will there be a Crucible III? That will depend entirely on how these urgent issues play out in the years ahead.

The Crucible II reports: *Seeding Solutions*

Those familiar with *People, Plants, and Patents* will recall that the book offered a summary of the major issues related to the ownership, conservation and exchange of plant germplasm. By and large, *Seeding Solutions,* Volume 1 has served a similar purpose. The volume, as succinctly as possible, brings readers up to date on changes — scientific, political, environmental — since the first report six years before. Readers benefit from viewpoint boxes summarizing the state of the debate as of late 1999, as well as surprising areas of agreement in the form of recommendations. From beginning to end, Volume 1 offers policy-makers a clear description of the facts, the fights and the fora relevant to genetic resources. Those new to these issues are also offered a clear picture of why germplasm is important and how it relates to trade negotiations, intellectual property disputes, and national and international food and health security.

Volume 2 of *Seeding Solutions* does *not* provide the answers to the issues posed in Volume 1. Indeed, some Crucible members might argue that Volume 2 demonstrates the absurdity of trying to solve sociopolitical issues identified in Volume 1 by applying purely legal mechanisms. Nevertheless, Crucible II's second volume does provide a range of technical legal options that national policy-makers can use to inform their own thinking about how to address issues identified in Volume 1. Readers are urged to study both volumes and to examine Volume 1 before launching into the more legalistic debates in Volume 2. Volume 2 is more technical in nature; readers will get much more from it if they have some previous exposure to the issues in the field. The Crucible II Group hopes that readers will find these two volumes helpful to their own understanding of the issues and in their own policy- and opinion-making activities.

The Crucible Group operates in good faith to produce best-effort non-consensus texts. Members of the group are individuals attending solely in their personal capacities. They have agreed to associate their names with this volume in the belief that the texts represent a helpful contribution to the global discourse on these issues. Members believe that the texts, in sum, accurately represent the current range of opinions and that these divergent viewpoints should be addressed. Probably, every member of the group is in strong disagreement with some general statements and many specific views provided in both volumes.

Volume 2 content and terminology: *articles, options, elements, viewpoint boxes* and *recommendations*

Four collections of optional legislative provisions for domestic laws make up the greatest part of this volume. The four collections concern access law, *sui generis* intellectual property protections for indigenous and local knowledge as well as those for plant varieties, and intellectual property protections for biological innovations. Each of these collections is divided into *articles*. Each article addresses a discrete issue that policy-makers must consider when making laws in these four areas. These articles are further subdivided into short, legislation-style provisions, each of which represents a different approach to the issue. We designate these provisions as either *options* or *elements*. Provisions that are mutually exclusive — that require the reader to choose one out of the menu of provisions presented — are *options*. Provisions that are not mutually exclusive, and that readers are free to accept or reject in combinations, are *elements*.

After each article, there is a comment section where we provide important background information such as the place of the more controversial options within continuing debates in the field, and their compliance (or lack thereof) with international legal agreements. In cases where disagreement among Crucible Group members about an article's significance is particularly strong, the parameters of those disagreements are set out in *viewpoint boxes*. In those instances when everyone in the group agrees about an important point, we make joint Crucible Group *recommendations*. Some individuals undoubtedly associate themselves more with some recommendations than others and, in many cases, individuals believe that the recommendations represent the lowest common denominator within the membership. Despite this, the group has avoided platitudes and striven for the highest achievable common denominator.

Readers of both Volumes 1 and 2 of *Seeding Solutions* will notice that there are fewer recommendations in Volume 2, and that their importance is not emphasized as much as in Volume 1. The reasons for this discrepancy are twofold. First, the point of this second volume is to present different legal options or approaches to policy issues — not necessarily to agree which options are best. Second, we have selected many of these options because they are controversial. In many cases, the tensions, regarding even relatively small issues, among these options represent microcosms of the larger debates in the field. These factors reduce both the importance and the likelihood of arriving at consensus-based recommendations in this volume.

The inclusion of any particular option or element in this text does not mean that it is endorsed by the Crucible Group. In fact, readers will see that this interpretation is impossible given that so many options or elements are mutually exclusive. If the Crucible Group does endorse a particular option or element, it does so explicitly in the form of a recommendation as noted above.

Some options presented in this volume may be incompatible with obligations that World Trade Organization (WTO) members have under the Trade-Related Aspects of Intellectual Property Rights (TRIPs) Agreement. These options have been identified throughout the text. Members of the WTO should carefully consider the implications of selecting such options. Members should be aware as well that some options presented here might not be consistent with the objectives of the Convention on Biological Diversity (CBD) or other international instruments. Where this inconsistency is either unquestionable or highly controversial, we have annotated it.

Acknowledgements

Members and management of the Crucible II Group

Crucible Group members participate in their personal capacities, but all have backgrounds in sectors or activities that inform their views on many of the issues taken up in this publication. Consequently, we have subdivided the signatories to the text of Volume 2 into the four broad sector categories in which they spend the majority of their professional time. We do not, however, provide details of individual institutional affiliation. The Crucible Group was served by a Management Committee; it was comprised of those whose names, included in the following list, are marked with an asterisk (*). In addition to overseeing the process, Management Committee members participated in writing, reviewing and deliberating over this text.

Signatories

Civil society organizations (including indigenous peoples' organizations):
Alejandro Argumedo, Margarita Flores, Glen Hearns, Dan Leskien, Atencio Lopez, Pat Mooney,* Andrew Mushita, Gurdial Singh Nijar, Olle Nordberg,* Rene Salazar, Priscilla Settee, Hope Shand, Vicky Tauli-Corpuz*

Private sector/industry:
Klaus Leisinger, Brian Lowry, Radha Ranganathan, P.V. Subba Rao, Tim Roberts, Reinhard von Broock

Academic:
Assiah Bensalah Alaoui, Carlos Correa, Don Duvick, Michael Flitner, Cary Fowler, Jaap Hardon, Francisco Martinez-Gomez, Michel Pimbert

Public sector:
Susan Bragdon,* Brian Davy,* Engsiang Lim, Chusa Gines,* Christine Grieder,* Geoffrey Hawtin, Michael Halewood,* Mita Manek, Norah Olembo, Theo van de Sande, Louise Sperling, Klaas Tamminga, Carl-Gustaf Thornström,* Joachim Voss,* Beate Weiskopf*

This second volume of *Seeding Solutions* complements Volume 1. Most of those who worked on Volume 1 also worked on Volume 2. We repeat here our profound thanks to all those whose work on Volume 1 laid the foundations for Volume 2. In specific connection with Volume 2, we would like to thank (or thank again) Dr Joachim Voss of IDRC who chaired the Crucible II proceedings for the first two years, and Dr Carl-Gustaf Thornström of Sida-SAREC, who took over the reins after Dr Voss was appointed Director General of CIAT.

*Members of the Crucible Group II Management Committee.

Wrestling all of the different points of view regarding options for national laws into this second volume required a great deal of time and patience on the part of everyone in the Crucible Group. A few members who served double or triple time in putting this volume together, however, deserve additional recognition. These individuals participated on one or both of (a) the *ad hoc* Legal Experts' Committee, which completed text revisions after each Crucible Group meeting, and (b) the Volume 2 Negotiating Committee, which undertook the volume's penultimate revision. Those who served on either or both of these committees are: Susan Bragdon, Michael Flitner, Michael Halewood, Glen Hearns, Dan Leskien, Engsiang Lim, Mita Manek, Pat Mooney, Radha Ranganathan, Tim Roberts and Vicky Tauli-Corpuz. Special thanks to Dan Leskien (who served on both) and did exceptional work on all the topics, in particular bringing his legal training to bear on the challenges of protecting biological innovations.

The group owes a particular debt to one member who has not been able to sign the document in its final form: Bernard Le Buanec. Despite his doubts about the practicability of the project, Bernard spent many hours poring over drafts. His criticisms in his areas of expertise have been wide-ranging and constructive, greatly benefiting the document as a whole.

Above all, the group owes an immeasurable debt to Michael Halewood, the Executive Secretary for the Crucible II Project. For over two years, Michael has lived and breathed Crucible: devising structures, creating drafts, writing, editing, leading discussions, organizing, encouraging and cajoling participants and dovetailing their contributions into a coherent whole. Without his enthusiasm and persistence, Volume 2 could never have been realized.

External reviewers, observers, editor, assistants and special guests

The Crucible Group recognizes its debt to the external reviewers of this volume. We would not have a text anywhere near the high quality that we present now without the advantage we gained through the comments and suggestions from: Professor Rosemary Coombe of the University of Toronto Law School, Professor Peter Drahos of the Queen Mary Intellectual Property Research Institute, Queen Mary and Westfield College, Tony LaViña, lawyer, Program Director, Biological Resources, World Resources Institute, and Allan McChesney, lawyer, international human rights law specialist.

We would also like to acknowledge the valuable contribution of José Esquinas-Alcazar, who participated as an observer.

The group also recognizes the contributions of Erin O'Manique, Stephen Langill and Kate Harrison at the IDRC Ottawa office in keeping the project moving over the last three years, providing everything from administrative assistance to editorial reviews and background research.

The group thanks Susan Bridges of BDK Communication for the final, comprehensive edit of this text.

Financial support, encouragement and participation

The Crucible Group wishes to thank the following organizations for their financial support and active participation in the meetings and discussions: the German Federal Ministry for Economic Co-operation and Development/Deutsche Gesellschaft für Technische Zusammenarbeit GmbH (BMZ/GTZ), Germany; the Canadian International Development Agency (CIDA), Canada; the Dag Hammarskjöld Foundation (DHF), Sweden; the International Development Research Centre (IDRC), Canada; the Swiss Agency for Development and Cooperation (SDC), Switzerland; and the Swedish International Development Cooperation Agency (Sida-SAREC), Sweden.

Finally, the group notes that KWS SAAT AG deserves a sincere thank-you for supporting the extra time and expertise that Dr Radha Ranganathan contributed to the project.

Introduction

Legal approaches to national genetic resource priorities

This second volume of *Seeding Solutions* follows naturally from the first. *Seeding Solutions,* Volume 1 analyzes technological and policy-related developments in the field of genetic resources over the six years following the first Crucible Group's publication, in 1994, of *People, Plants, and Patents.* Volume 1 identifies three areas of policy development where there are contentious 'outstanding issues'. Volume 2 is divided into three *topics* corresponding generally to those outstanding issues. Each of the three topics is comprised of options for national laws that policy-makers can consider to address these issues.

The first outstanding issue identified in Volume 1 is the need to ensure the most open and equitable possible flow of genetic resources between farmers and other rural people, and researchers. Topic 1 of Volume 2 begins with a general discussion of a range of domestic legal options that policy-makers can consider to regulate access and exchange of genetic resources. This is followed by a more in-depth analysis of one of the approaches identified in the general discussion: a national law to create a framework for bilateral negotiations between parties seeking access to biological resources and those with legal rights over those resources.

The second outstanding issue identified in Volume 1 is the need to protect, promote and conserve the knowledge of indigenous and local communities in ways that ensure those communities' full participation in germplasm conservation and enhancement. In Topic 2 of Volume 2, we identify a range of options that could be implemented at a national level to protect, conserve and promote indigenous and local knowledge. We follow up with a more detailed analysis of legal options for the creation of domestic *sui generis* intellectual property laws for indigenous and local knowledge.

The third outstanding issue identified in Volume 1 is the need to encourage innovative research on genetic resources for the benefit of present and future generations. Topic 3 of Volume 2 commences with a general discussion of policies domestic governments could pursue to encourage such innovation. In two subsequent sections of Topic 3, we analyze options for two different kinds of intellectual property laws: plant variety protection laws and patent laws.

In the end, out of all of the legal approaches identified in this volume, we dedicate what may seem a disproportionate amount of attention to the analysis of bilaterally oriented domestic access laws (in Topic 1) and intellectual property laws (in Topics 2 and 3). This does not signify that Crucible Group members jointly endorse these legal approaches as the best means to address the outstanding issues. In fact, many members of the Crucible Group argue that bilaterally oriented national access and intellectual property laws will ultimately have a negative impact on the resolution of these very issues.

Instead, our rationale for this focus is that, although these laws have been the subject of so much recent controversy, there has been little progress in public discourse on their ultimate utility and technical feasibility. Despite disagreement among Crucible members about these laws, we all recognize the value of a multistakeholder, non-consensus–based group working through and disentangling the associated issues.

At the end of the day, we know we have not 'solved' the problems currently facing national policy-makers regarding domestic access and intellectual property laws, to say nothing of all of the other policy initiatives we have identified in the survey sections of this volume. However, by charting the range of options available to policy-makers and annotating these options with criticisms from the different perspectives represented within the group, we hope to facilitate the continuing debate.

This volume demonstrates, among other things, that most of the laws and legal principles currently available to policy-makers are ill suited to the goals of maximizing access, exchange and innovative uses of genetic resources. The field of genetic resources engages an extraordinarily diverse and novel array of interests, actors and issues. A great deal of this volume is dedicated to trying to make contract-based, bilaterally oriented access laws and intellectual property laws 'fit' this new field. It is not surprising that they make a rather clumsy fit in many cases.

We want to warn readers away from the notion that any of the options set out in Volume 2 constitute complete, easy, technical solutions to any of the problems raised in Volume 1. Attractively controversial talk about access and *sui generis* intellectual property laws should not lead policy-makers to conclude that they represent genetic resource policy panaceas. For example, it is important to assess the way in which many countries are approaching access legislation, and to think about how it may be structured to account for the particular nature of plant genetic resources for food and agriculture (PGRFA). If systems are being instituted that inadvertently put a chill on the exchange of PGRFA, countries may with one hand undercut important activities relevant to national interests — such as food security — supported by the other. If policy-makers are seriously considering national access and *sui generis* intellectual property laws as means to forward their national policy objectives, they must also continue to examine other, potentially easier and more fruitful options such as those set out in the survey sections of each topic in this volume. In most — perhaps all — cases, policy-makers may find it helpful to pursue a number of very different policy initiatives in order to achieve their ultimate goals.

Finally, a study such as this cannot be conducted in a policy vacuum. Consequently, we have made every effort to situate the options we include in this volume in the larger context of international, national and community law, and policy developments and debates. To that end, we have annotated the options with reference to:

1 international instruments such as International Union for the Protection of New Varieties of Plants (UPOV) Agreements, the International Labour Organisation's Convention 169 concerning Indigenous and Tribal Peoples in Independent Countries (ILO 169), TRIPs, CBD, the Draft Declaration on the Rights of Indigenous Peoples, and the Organization of African Unity (OAU)'s draft model legislation on Community Rights and Access;
2 national laws and policy initiatives such as the Philippines' Indigenous Peoples' Rights Act and Peru's draft law on Traditional Knowledge, and several countries' patent and plant variety protection laws;
3 peoples' declarations such as the Mataatua Declaration on Cultural and Intellectual Property Rights of Indigenous Peoples and the Indigenous Peoples' Earth Charter;
4 voluntary organizational protocols such as the Food and Agriculture Organization of the United Nations' Code of Conduct (FAO-COC); and
5 relevant secondary research.

We have not attempted to provide options that invariably comply with these precedents. Instead, our approach has been to note that a particular provision might be, or definitely would be, in contravention of a binding international agreement such as TRIPs.

List of viewpoint boxes

Topic 1

Options for national laws to regulate access to biological resources

Contents

Section 1

Introduction

The discussion of access and exchange of genetic resources in the 'Outstanding issues' section of *Seeding Solutions,* Volume 1 dealt as much with international negotiations and agreements as it did with national issues, or more. In this volume, the Crucible Group focuses exclusively on national legal options.

We do not get into analyses of international, multilateral access negotiations and agreements such as the negotiations for the International Undertaking on Plant Genetic Resources for Food and Agriculture (IU on PGRFA).[1] We do, however, include options to exclude material from the scope of national access laws where it might eventually become the subject of international access agreements. To this end, the 30th Conference of the FAO "stressed the importance for countries that are developing relevant legislation to do so in such a way that would enable them to take into account and allow for the elements of this new international agreement." This was reiterated by the Conference of the Parties to the Convention on Biological Diversity (through decision V/26(A)(7)), which requested Parties to the CBD, when "developing national legislation on access", to "take into account and allow for the development of a multilateral system to facilitate access and benefit-sharing in the context of the International Undertaking on Plant Genetic Resources". The multilateral system (MLS) will cover a list of crops drawn up on the basis of two criteria: the interdependence of regions and countries with regard to the crop, and the significance of the crop for food security. Because the free flow of PGRFA underpins food security for all, the establishment of an MLS is intended to minimize transaction costs, obviate the need to track individual accessions and ensure expeditious access in accordance with applicable property regimes.

Nor do we analyze legal options for countries into which accessed materials are imported. It is possible, for example, to imagine an importing country passing customs laws to require proof that the resource-supplying country's access laws were obeyed before allowing material into the importing country. We include those options in Topic 3 as conditions for the grant of plant variety protection rights and patents.[2] In this topic, we focus on the supply end of the access relationship.

There appears to be a broad range of opinion concerning what national access laws can be made to do, and by extension, the kinds of legal mechanisms they should include. There is also terminological imprecision regarding access laws and their relationship to other kinds of laws that might affect the same biological resources. For example, it is not uncommon to hear people referring to 'access laws', '*sui generis* laws' and 'access laws to protect indigenous knowledge' all within the same conversation, as though these terms were synonymous. But each of them engages a different combination of underlying legal categories and suggests a different legal response. In the Appendix to this volume, we have included a discussion of the relationship of access law to intellectual property laws, the legal categories that they intersect, and the manner in which they can overlap.[3] We suggest that readers refer to that discussion as they read Topic 1.

The members of the Crucible Group are divided about the usefulness and effectiveness of national access laws. Many hold the view that the whole enterprise of creating national access laws is built upon unrealistic economic expectations. They do not believe that access laws can be made to turn a significant profit. Others are concerned that emphasis on creating laws to facilitate *ad hoc*, bilateral deal-making is distracting critical attention from efforts to create more important international agreements for multilateral access to, and exchange of, genetic resources. As a consequence, Southern countries that should form a unified front to conduct international negotiations for multilateral access are instead fractured, sometimes taking mutually disadvantageous positions. Consequently, some critics conclude that unrealistic expectations regarding the benefits of bilateral access deals are ultimately disempowering Southern countries. To the extent that the CBD encourages these expectations and the concomitant diversion of resources to bilateral-access deal-making, they argue that it too contributes to the disempowerment of these countries.

On the other hand, some Crucible Group members argue that well-constructed national access laws can compel significant benefit-sharing where none existed before. They contend that even if the benefits of such regulations turn out to be smaller than originally expected, it is still definitely worthwhile to have them in place.

The Crucible Group does not attempt to resolve these issues. Ultimately, it is up to national policy-makers and advocates to decide if they want to create national access laws, and what elements they want to include in them. We hope that the analysis in this topic makes that task somewhat less daunting.

Section 2

Options for national access laws

Contents

Section 2

Options for national access laws

Article 1: Purpose

The purpose of a national law to regulate access to biological resources is:

Element 1
to ensure national sovereignty over genetic resources in accordance with national and international law, including the implementation of Articles 15 and 16.3 of the Convention on Biological Diversity (CBD), and the International Undertaking on Plant Genetic Resources for Food and Agriculture (IU on PGRFA)

Element 2
to contribute to food security and health

Element 3
to contribute to the conservation and sustainable use of biological resources

Element 4
to promote the equitable sharing of benefits derived from the use of genetic resources

Element 5
to recognize the rights and needs of indigenous and local communities and farmers, and those who manage wild and domesticated genetic resources

Element 6
to facilitate access in order to promote development, use and deployment of resources in a manner relevant to national needs and objectives

Element 7
to encourage the development and transfer of appropriate technologies to the

relevant sectors and communities of the resource-supplying country [for the sustainable use of biological resources]

Element 8
to enforce compensation to the supplying country and/or community when genetic resources are taken out of that country and/or community

Element 9
to regulate the process by which collectors may legally obtain genetic resources

Commentary

National legislation is not required to have a 'purpose' section. In cases where legislative texts are vague (often because of political disagreements among the legislators who drafted it), purpose sections are valuable interpretive tools. Reference to the overall purpose of a piece of legislation can help policy-makers to choose among different possible interpretations of ambiguous text. Certainly not all of the possible purposes of such legislation are included here. Ultimately, it is up to national policy-makers and policy advocates to determine why the country needs an access law in the first place (or if it needs it).

Element 1 refers to the two international legal agreements that are most directly relevant to the creation of national access legislation. While the CBD evidently sanctions the kind of national access law we are considering here — one primarily oriented towards facilitating bilateral deal-making — it does not *require* countries to establish such laws. The CBD recognizes that states have a sovereign right to establish the terms of access to their own resources. At the same time, however, the CBD implicitly limits this right by encouraging states to set conditions that facilitate, rather than hinder, that access. There are many ways that countries can facilitate access without passing this kind of national legislation. Having no law whatsoever, and 'throwing the door open' to any and all parties, would be one such means. Entering into multilateral agreements to establish global- and regional-access resource-based networks to promote and facilitate exchange is another. The IU on PGRFA currently being negotiated under the auspices of the Food and Agriculture Organization of the United Nations (FAO)'s Commission on Genetic Resources for Food and Agriculture (CGRFA) is an example of a potential multilateral system intended to facilitate access for identified crops. Many, if not most, countries already belong to crop-based networks.[4] One objective of these networks is the multilateral exchange of material. While a multilateral agreement of the nature of the IU on PGRFA tends to minimize the conditions of access, it is nonetheless an expression of sovereignty on the part of states choosing to

Viewpoint box 1: Do national access laws facilitate bilateral deal-making?

They simultaneously promote food security, biodiversity conservation, equitable benefit-sharing and technology transfers.

The recognition in the CBD that countries have the sovereign right to regulate access to the genetic resources within their borders was a truly significant move. It has finally given developing countries the legal and economic leverage they need to effectively regulate access to their resources. This is particularly important for countries with many endemic species. Now, imaginative, activist governments of countries with a high degree of biological diversity can coordinate agreements involving themselves, their local populations and foreign researchers that simultaneously (a) justify conserving large tracts of 'traditional' ecosystems, (b) include technology transfers and training for local people, and (c) enforce monetary compensation.

The confidence that such laws provide developing countries naturally salves the feeling that they are being exploited and counteracts the protectionist 'closed-door' ethic of severely restricting access. In this way, national access laws contribute to the global movement of genetic resources.

They may be useful for endemic, localized genetic resources, but will work against the interests of most countries when it comes to plant genetic resources for food and agriculture.

The exchange and use of agrobiodiversity[5] is fundamental to agricultural production and food security. Since the beginning of agriculture some ten thousand years ago, human beings have relied on the genetic diversity of plants to develop a wide range of genetically diverse crops and other useful species that have enhanced our survival.

Today, the agriculture of virtually all countries depends on a supply of resources from other parts of the world. Sub-Saharan Africa, for example, is estimated to be 87% dependent on other parts of the world for the plant genetic resources it needs. Even the countries considered richest in biodiversity are dependent on plant genetic resources from other parts of the world. Access laws that do not distinguish among different types of genetic resources risk putting a chill on the kind of exchange that forms the foundation of a country's well-being.

They are an expression of geopolitical rhetoric and nothing more.

The whole idea of boosting the profile of regulated deal-making for the supply of genetic resources from South to North (through the CBD) is a politically driven smokescreen which very cleverly appears to address geopolitical inequities, but in fact does almost nothing. Rhetorical dedication to international benefit-sharing and promotion of technology transfers allows overpowered developing-country delegates to return home from international negotiations without having to admit their lack of progress to constituents starving for signs of improvement. National laws purporting to put those principles into action just take the ruse one step farther: conspicuous (legal) displays of putting the nasty North back in its place. But they don't actually do anything. Biological resources are already so widely spread throughout the world that it would be extremely rare for anyone who badly needed a genetic resource (enough to pay decently for it) to have to go to the country of origin to obtain it. The fact that there are a number of potential sources for a particular genetic resource makes it practically impossible for a supplier to bargain for a decent price.

Foreigners may enter into access agreements, not because of the strength of the legal position of the supplying country, but because it looks great for public relations. Under such circumstances, the terms of payment will be negligible.

Finally, if they actually paid off, why aren't any Northern countries passing national access laws? The truth is, Northern countries don't need the only real benefit these laws can deliver: political points for making a rhetorical stand against North–South exploitation.

enter into it. In Article 3, Element 7, we address the possibility of exempting designated material from the bilateral deal-making that these options are primarily intended to create, in order to 'free them up' to be subject to multilateral agreements.

Elements 2 through 9 do not refer to any particular international laws. Instead, they refer to substantive objectives or purposes that national law-makers may seek to advance through the creation of national access laws of this nature. Each of the purposes listed is successively less ambitious. Crucible members hold very different opinions of the usefulness and effect of this kind of legislation, as is reflected in Viewpoint box 1.

Article 2: Scope

This Law shall apply to access to:

Paragraph 1: Types of resources

Option 1
wild species of flora and fauna

Option 2
domesticated and cultivated species of flora and fauna

Option 3
any [genetic] [biological] resources

Paragraph 2: Source of resources

obtained from:

Option 1
in situ sources

Option 2
ex situ sources

Option 3
in situ or *ex situ* sources

Paragraph 3: Geographical locales

located in:

Element 1
this country's territory within its internationally recognized borders

Element 2
and its territorial waters

Element 3
and the various maritime zones adjacent to them

Paragraph 4: Country of origin/source country

Option 0
no provision [i.e., the country may regulate access to any biological resource within its jurisdiction, despite the fact that it may have, at some point, come from another country]

Option 1
and for which this country is the country of origin.

Paragraph 5: Associated information

This Law also applies to:

Option 0
no provision [i.e., the law is silent on the issue of whether or not associated information can be included; in the absence of explicit exclusion of associated information, having no provision would probably be interpreted to permit parties to include associated information in access deals passed pursuant to the law]

Option 1
information associated with the resources supplied pursuant to this Law.

Commentary

This article defines the scope of the access law. It outlines which activities are subject to the access law. The article should be read in connection with Article 4, which defines the term 'access' as such. It is useful to keep the scope of the law broad to encompass all relevant activities. That said, the law should not extend to such things as agricultural crops sold for consumption as food, or the sale of trees for use as lumber. Options for exclusions of this nature are considered in Article 3, below.

Paragraph 1 (Types of resources): All three of these options are assumed to refer to a combination of plants, animals, microbes, fish, insects, and so on.

We have chosen not to break down the types of resources that might be covered into such classes. Of course, a national law could do that, although it would be an unlikely practice.

Paragraph 1, Option 3: According to Article 2 of the CBD, 'genetic resources', are any material of plant, animal, microbial or other origin containing functional units of heredity (of actual or potential value). This definition is limited, as it does not include such things as secondary products but rather focuses on units of heredity. One may consider the term 'biological resources' to be broader, as it might include, apart from genetic resources, also secondary naturally occurring products such as naturally occurring molecules or combinations or mixtures of molecules including extracts from living or dead naturally occurring organic matter. One may establish these definitions in the law or, alternatively, simply regulate that genetic resources and derivatives shall fall under the law.

Paragraph 4 (Country of origin/source country), Option 0 would see resources regulated in the same way whether or not the implementing country is the country of origin.

Paragraph 4, Option 1 would limit the scope of the law to resources for which the country is the country of origin.

There is a middle ground. The implementing country could regulate access to both categories of resources, but in different ways. For example, regarding resources for which the implementing country is not the country of origin, the law could require both (a) prior informed consent (PIC) from specified consent-granting parties of the implementing country, and (b) PIC from specified consent-granting parties from the country of origin of those resources.

The term 'country of origin' is problematic when applied to plant genetic resources for food and agriculture (PGRFA). Article 2 of the CBD states that the country of origin of genetic resources means "the country which possesses those genetic resources in *in situ* conditions." The CBD further defines '*in situ* conditions for domestic or cultivated species' as those "surroundings where they have developed their distinctive properties." The Convention does not, however, define 'distinctive properties'.

Owing to the history of agriculture, many important crops are widely distributed around the world and have developed unique adaptive characteristics in regions far from their origin. PGRFA have been moving around the globe for thousands of years, encountering and adapting to new conditions. Consequently, for most of these resources, it will be extremely difficult — if not impossible — to determine their country of origin as defined by the CBD. In looking at individual crops, both genetic diversity in general, as well as specific characteristics, are generally found in more than one country. Hence,

Viewpoint box 2: Should 'associated information' be included in the scope of national access laws?

No.

'Associated information' should not be included within the scope of national access laws. Its intangible nature makes it qualitatively different from tangible genetic resources. Following the same approach for both would lead to confusion and would probably be unimplementable. Associated information should be covered by intellectual property laws.

Depending on the manner in which the supplying party participates in the collection and delivery of resources, it only makes sense to allow parties to include 'associated information' in bio-access deals.

'Associated information' can include an almost infinite range of information. It can be relatively simple, such as details about the location and environment from which a sample is collected, or observations about the life cycle of the resource. Alternatively, the associated information could be quite complex, including details based on generations within a community working with the resource, or information gathered from bio-assays conducted on the material (if the access contract includes the term that the supplying party will conduct a preliminary bio-assay before delivering resource samples to the applicant). To preclude parties negotiating access agreements from including the supply of, and compensation for, such information within access agreements would be to limit the creative potential of such relationships to deliver benefits to both sides.

That said, as long as the law does not explicitly preclude the supply of associated information, it will not make any difference whether any mention of it is actually included in the law or in any agreements struck pursuant to the law. Parties to the agreements need only satisfy themselves that a deal is fair and worth entering into. Whether or not the 'value added' by local people, who may provide information about where to find the resource or about its medicinal properties, is recognized in the agreement as associated information is of no significance as long as the parties to the agreement are mutually satisfied with its terms. The agreement could be silent with respect to associated information but, nonetheless, compensate a local community well enough that its members would be willing to provide it.

Parties who hold associated information in a form protected by intellectual property law (bearing in mind that intellectual property does not protect information *per se*) can negotiate licences with parties who want to use that information or the resource in which the information is embodied (e.g., a plant variety or biological innovation). Whether or not an access agreement includes licensing provisions for the supply of protected intellectual property should be entirely up to the parties making the deal.

Some of the associated information may already be protected by intellectual property laws (e.g., patents, plant variety protection or *sui generis* laws for indigenous and local knowledge). This situation could be dealt with in various ways. For example, knowledge that is subject to intellectual property protections could be exempt from the scope of the access law, thereby requiring access-seekers to go directly (and only) to the owner of the intellectual property right in that knowledge. Alternatively, the access law could require PIC from both the intellectual property owner and the government. (We consider these options in Article 6, below.) The point here is that the existence of intellectual property rights over associated knowledge does not dictate whether that associated knowledge should be included or excluded from the ambit of an access law. It does, however, have to be taken into consideration, and some provision would have to be included in the access law to inform how such associated knowledge would be dealt with.

the country where a crop variety or farmers' variety is collected may not correspond to the CBD-defined country of origin of the species, the variety, or any particular characteristic it displays. Genebanks with good historical data may be able to identify where an accession was collected; this is very different from identifying where it acquired its 'distinctive properties'. Even in cases where the multiplicity of source countries can be identified, it is unlikely that substantial revenue or other benefits will accrue to them. That would require a whole series of bilateral arrangements with a number of different countries, each of them a party to benefit-sharing negotiations.

Paragraph 5 (Associated information), Option 1 makes it explicit that associated information could be included in an access law. Including such a term would clear up any ambiguity that would exist if the law were silent on the issue (as in Option 0).

In most commentary regarding access laws, the term 'associated knowledge' is used. 'Associated information' is less ambiguous, and avoids many difficulties that arise because there is no shared understanding of the meaning of the word *knowledge*. (For more on the vagaries of the definition of 'indigenous and local knowledge', see Topic 2, Section 1, subsection B.4, 'Legal imprecision', below.)

Article 3: Exemptions

Option 0

no provision [i.e., there are no exemptions from the blanket provisions created in Article 2]

Option 1

This Law shall not apply to access to material otherwise covered by this Law when that access is:

Customary use

Element 1

part of the customary practices of indigenous and local peoples

Commodities

Element 2

for the purpose of direct use or consumption

Material of human origin

Element 3

for any components or substances derived from the human body, or access to the human body itself

Farming

Element 4

for farming purposes

Private use

Element 5

for non-commercial private and personal use

Research

Element 6

for non-commercial scientific research purposes

Exempted species

Element 7

for PGRFA designated by the competent authority. The competent authority may designate PGRFA, access to which will be governed by special regulations or conditions such as those established by PGRFA networks, multilateral systems of exchange such as the IU on PGRFA[6] or other facilitated PGRFA access arrangements reflecting reciprocity among their parties and which [the implementing country] has joined

Private property

Element 8

to resources that are located on or in private real property, personal property or resources protected by intellectual property laws

Associated information

Element 9

This Law will not apply to information associated with resources covered by this Law.

Commentary

Exemptions are necessary to make the law practical and implementable. Because genetic and biological resources are very broadly defined, access laws, if not narrowed down, can cover many activities and result in absurd applications. For example, in the case of taxonomic research, they could halt or slow necessary ecological research with unintended negative environmental consequences.

Option 1 offers a whole range of possible exemptions from the scope of this law. Existing access laws have often been criticized as overbureaucratic and as creating serious obstructions to germplasm exchange. This article offers the opportunity to shape the law's scope so that it focuses on those resources and activities from which a country expects the most in terms of benefit-sharing agreements.

Element 2: It would certainly make sense to exclude commodities from the scope of the law since it could otherwise severely hamper trade in biologically based products such as food and timber.

Element 3: There is general agreement that the CBD does not apply to human genes, although the terms used by the Convention technically would allow human genes to be treated like any other genetic resource. The legal questions arising in the context of human genomics might be sufficiently different from those addressed by this kind of access law to make it worthwhile to create a separate legal regime regarding human material.

Element 6: Countries may wish to distinguish between access for commercial purposes and for research purposes. It should be noted, however, that the line between research and commercial applications has become very thin. In many cases, public research projects are funded by private-sector companies. In return, these companies usually get the option to use and commercialize the research results.

Element 7: This provision envisages the completion of negotiations at the CGRFA and the creation of a mutually agreed list of varieties and species that would be subject to special rules of access and exchange. Either this legislation would not apply to these resources, or it would include a special section wherein the domestic component of the multilateral access regime established pursuant to the IU on PGRFA was implemented. We include the idea here simply to provide readers with an example of how the results of the revised IU on PGRFA negotiations could fit within the context of bilaterally oriented access legislation. It is unlikely to be sufficient to simply exclude those plant genetic resources covered by a potential IU on PGRFA. The negotiations revising the IU indicate that its scope is likely to be some subset of PGRFA. Those drafting access legislation will need to consider how to promote the exchange of PGRFA — if it is determined that this is in the national interest — not only within the scope of the IU on PGRFA, but outside it as well.

Element 8: According to this element, private property owners would have the right to conclude a proposed access agreement without obtaining government permission. This approach could have the effect of encouraging access-seeking parties to seek out private property owners as a means of avoiding the application of the access law. Whether or not this resulted would depend upon how onerous the access law was and how difficult to deal with private property owners proved to be.

It is important to consider the legal characterization of the lands and territories of indigenous peoples and local communities. Sometimes such lands are characterized as the private property of the people in question, and so would fit within this exemption. In other cases, these lands are characterized as being owned by the government, with the indigenous peoples enjoying a limited (non-proprietary) right to use those lands for designated purposes. In the latter case, would the exemption apply, with the indigenous peoples being treated like private property owners?

The effect of not exempting resources located on privately owned property would be that access-seekers would have to get the permission of the competent authority as well as the private property owner. This option corresponds to Article 6, Option 2, below.

Element 9 is the antithesis of Article 2, Paragraph 5, Option 1, which explicitly states that associated information may be included in access agreements created pursuant to this law.

Part Two: Definitions

Article 4: Access

'Access' means:

Option 1

obtaining, collecting, utilizing and/or exporting material [derivatives] [associated information] covered by this Law.

Article 5: Applicant

'Applicant' means:

Option 1

any person (natural or legal) seeking to gain access.

Option 2

any non-national person (natural or legal) not having their place of business in this country seeking to gain access.

Commentary

Option 1 treats national and non-national parties seeking access in the same way.

Option 2 makes it such that only foreigners would have to apply for access. Domestic parties would be exempt. Foreign-based companies with branch plants are somewhere between the two; policy-makers could decide to require them to apply for access, or they could be exempt as well. Lawmakers might embrace this form of discrimination in order to give locally based industries an advantage. The rights and obligations envisaged by the CBD are primarily of international character, thereby maintaining the possibility of treating nationals differently from foreigners. However, this form of discrimination may violate the World Trade Organization (WTO)'s national treatment requirements, which stipulate that foreign parties be treated the same as national parties. Quite apart from the state of international law, however, policy-makers may decide to require domestic and foreign parties with branches in the country to adhere to the same standards as foreigners with no presence in the country, particularly with respect to such issues as PIC and indigenous and local community rights. This could act to correct unfair practices among domestic parties. In this way, domestic research institutions and commercial enterprises would be equally duty bound to respect, for example, the rights of domestic local communities. Furthermore, in many cases, foreign parties work with local partners. Exempting local partners from access legislation could provide a big loophole for foreign parties to exploit.

Part Three: Conditions of access

Article 6: Parties authorized to grant consent

Access covered by this Law shall ultimately be subject to the prior informed consent of:

Option 1
the competent authority of this country. [i.e., only the appointed competent authority would have the right to say 'yes' or 'no' to an application; there is no provision to require the prior informed consent of private parties or local and indigenous communities]

Option 2
1 the competent authority of this country, and
2 the local community or person with any exclusive right or interest in the material [or derivatives] [or associated information].

Option 3
the local community or person with any exclusive right or interest in the material

[or derivatives] [or associated information]. [i.e., only the local community, and there is no need to obtain the approval of any government authority]

Commentary

Option 1: Although far from satisfactory from the community point of view, the CBD does not require that local communities consent to 'access' proposals.

Option 2: The Philippines' EO 247[7] and related regulations require the consent of both the national government and the involved local communities. Certainly, nothing in the CBD prohibits setting up a national system that requires the PIC of communities. Indeed, Article 8(j) implies that the approval and involvement of local and indigenous communities should be sought, at least concerning access to knowledge. How to decide who is responsible for considering access applications within indigenous and local communities is a related issue. It is possible to imagine a scenario wherein domestic laws governing the relationship of the national government and indigenous or local peoples might dictate which person, or which body within the community, should make this decision. A better approach would be to leave this decision to indigenous and local communities themselves.

There are several different ways to structure the relationship between the three principle parties to an access agreement: the access-seeker, the supplier and the government. One common approach is to have a contract between the access-seeker and the domestic supplier. This contract would embody, among other things, the PIC of the supplier to the terms of access. A second agreement

Viewpoint box 3: Who should have the right to say 'yes' or 'no' to a proposed access agreement: the government alone, the supplying parties alone (when the supplier is not the government), or both?

Suppliers alone.

It is paternalistic to make private parties and/or indigenous and local communities get the approval of government before they can make a deal. If, through land law, personal property law, intellectual property law, or the recognition of self-governing rights, they would otherwise have exclusive control over the use of the resource, it is absolutely unnecessary (and unfair) to 'deal in' the government through the creation of a national access law.

Both.

Respecting basic human rights requires that communities and private parties have a direct say regarding whether or not access should be allowed. Compelling national interests (the economic value of genetic resources and the need to develop science and technology) suggest that society at large have a stake in access issues. Consequently, governments must also give consent.

would then be forged between the supplier and the government. This contract would, among other things, embody the PIC of the government. This kind of 'two-contract' approach to structuring an access agreement has been followed in a number of access agreements in various countries including Costa Rica and Brazil.

Another method is to have the access-seeking party approach the government, which would then seek the PIC of the supplying parties. Following this format, there would be a single agreement with an annexed document providing proof of the PIC of the supplier. This is the approach set up in the Philippines' EO 247.

In either case, the supplying party ultimately has a right to grant or not grant its authorization. The two-contract approach, however, appears to envisage more independence on the part of the supplier, at least procedurally, and more interaction with the applicant in early stages of negotiations.

Option 3: Under this option, only the person, community or other legal entity holding an exclusive interest or right, for example, in the land where the resources are located, has the right to consider access applications. This is evidently the Canadian and US position. Even though Canada does not have any form of centralized body to consider applications for access pursuant to the CBD, Canada has taken the position that it is already in compliance with the CBD by virtue of the laws it already has in place.

Article 7: Application for access permit

To enable the permit-issuing authority [and the community or person] to decide to grant or refuse a permit, the applicant should, when submitting an application to access material [or derivatives] [associated information] covered by this Law:

Element 1

provide a description of the applicant, including its legal status, place of residence and a list of all other entities and individuals that will be involved in the access activity, along with their respective responsibilities

Element 2

describe the applicant's technical and financial capability to conduct the access activity and previous biological resource collection activities within [the implementing country] and elsewhere

Element 3

identify, as far as possible, the biological resources it is seeking

Element 4
identify the purpose of the access activity

Element 5
state whether there is any intention to commercialize as a result of the activity

Element 6
identify the limits of the geographical location in which the applicant wishes to conduct research-related activities — including provisional route, estimated timing of expedition, types of material to be collected, species and quantities — and identify the methods of collection (sample, harvest methods, storage methods) intended to be undertaken by the applicant, either alone, or in association with the parties providing access

Element 7
describe the anticipated benefits and how they will be distributed if the resource or any product derived from it or related to it is commercialized as a result of the access activity

Element 8
identify the nature of the legal rights the applicant may seek over the collected resources, derivatives of the collected resources, and innovations derived from those resources, including any intellectual property rights, trade secrets and marketing rights

Element 9
provide information about existing or proposed contracts between the applicant and any third party relating to the use of any information and products resulting from the access activity

Element 10
identify the methods by which the applicant will transport the collected resources and how the applicant will ensure the safe use of those resources once transported back to the applicant's home country or institution

Element 11
provide information about the kind of assistance that may be required to facilitate the mission's success

Element 12
identify the individuals and or communities that the applicant wishes to associate with concerning its related activities

Element 13
indicate plans for cooperation with national scholars, scientists, students, civil society organizations and others who may assist with or benefit from participation in the field mission or its follow-up activities

Element 14
list, so far as it is known, the national and foreign curators to whom the germplasm and information is intended to be distributed on completion of the mission

Element 15
demonstrate that the collection/access activity will not have an adverse environmental impact

Element 16
provide translation of all the information into the official language of this country and local languages where appropriate

Commentary

This article is a combination of terms taken from various material transfer agreements: the Philippines' EO 247, emerging draft bio-access legislation and the FAO Code of Conduct for Plant Germplasm Collecting and Transfer (FAO-COC). The terminology has been adapted to encompass biological resources. The structure follows that of the FAO-COC.

We have listed a wide range of options. Including all of them might make the laws too cumbersome. Policy-makers must be selective in what elements they choose.

Element 4 could be problematic, inasmuch as access-seeking parties might not actually know what the collected material might eventually be used for. They would not necessarily want to agree in advance not to put the material to unforeseen uses. Various consequences could follow from the inclusion of this section in the law. If, two years after collecting the material, the collector discovers a new use for it, the collector could be required to return to the supplier for additional permission.

Element 8 requires parties to predetermine their legal rights with regard to collected material and any downstream innovations using that material. It will be important to determine the extent of legal rights of all parties. How far should these rights extend? If there are secondary commercial products, what are the rights of the country of location with respect to them? These questions lead back to the concept of direct derivation and should be compatible with provisions in plant variety protection legislation. (See Topic 3, Section 2, Article 5, 'Essentially derived varieties', below.)

Element 13: The collector is likely to interact with a variety of groups including national research institutions, academics, government organizations, civil society organizations and local communities. The nature of the collection activity will provide opportunities for nations to be engaged in the activity, and these opportunities and the nature of the relationship should be clarified in the application.

Element 15: This addresses *inter alia* a biosafety-related concern regarding movement of biological materials to areas where they are not endemic.

Article 8: Conditions for the grant of access

Option 0

no provision [i.e., there are no mandatory terms applicants must agree to in order to obtain access]

Option 1

Access shall be granted if the following minimum requirements are satisfied:

Element 1

the indigenous and/or local community or individual concerned has given prior informed consent

Element 2

the applicant agrees to adhere to a limit on the quantity, and specifications of the quality, of the biological resource that the collector will obtain and/or export

Element 3

the applicant guarantees to deposit duplicates of each specimen of the resource or the records of community innovation or knowledge collected with the designated authority and, if so required, with local community organizations

Element 4

the applicant agrees to inform the competent authority and the concerned local community of all findings from research and development on the resource

Element 5
the applicant agrees not to transfer the resources accessed, or any derivatives or associated information, to any third party without the authorization of the competent authority and the local community concerned [unless the third party agrees to observe the conditions originally agreed to by the applicant]

Element 6
the applicant agrees to obtain the permission of consent-granting parties prior to applying for [a patent] [intellectual property rights] relating to the biological resources or any derivatives or for a patent relating to an invention based on associated information covered by the Law

Element 7
the applicant agrees to benefit-sharing conditions negotiated in accordance with Article 9, 'Conditions for benefit-sharing'

Element 8
the applicant agrees to submit, to the competent authority, a regular status report of research and development on the resource concerned, and, where the biological resource is to be collected in large quantities, on the ecological state of the areas

Element 9
in order not to increase the risk of genetic erosion, the acquisition of germplasm must not deplete the populations of the farmers' planting stocks or wild species or remove significant genetic variation from the local gene pool

Element 10
the applicant agrees to conduct an environmental impact assessment

Element 11
the applicant agrees to abide by the relevant laws of the country and to respect local customs, traditions and values, and property rights and has demonstrated a sense of gratitude towards indigenous and/or local communities, especially if use will be made of local knowledge about the characteristics and value of germplasm. Collectors should respond to their requests for information, germplasm or assistance, to the extent feasible

Element 12
the applicant agrees to inform the local communities and farmers concerned about how and where they could request and obtain samples of the collected germplasm. If requested, the collector will provide duplicate samples to them

Element 13

whenever germplasm is collected, the collector systematically records the passport data and describes in detail the plant population, its diversity, habitat and ecology, so as to provide curators and users of germplasm with an understanding of its original context. For this purpose, local knowledge about the resources (including observations on environmental adaptation and local methods and technologies of preparing and using the plant) should also be documented; photographs may be of special value

Element 14

the applicant agrees to make available any technologies derived from material collected in this country for local use without restrictions [at a rate to be agreed upon by the collector and the competent authority]

Element 15

in addition to all other requirements, the issuing [competent] authority [and the local community or person] may where necessary or appropriate require the collector to comply with additional requirements

Commentary

Element 1: To make the inclusion of PIC in a national law meaningful, the law must specify processes and minimum substantive conditions under which that consent can be obtained and granted. Those procedures and substantive conditions must work together to ensure transparency. For example, the laws should specify how public consultations should be undertaken, how much information applicants must disclose, who should be notified about applications for access, when it would be appropriate to refer to the customary laws of the communities from whom PIC is being sought, and so on.

Elements 5 and 6 address the issue of derivatives. In many cases, researchers are less interested in the resources or the raw materials themselves than in the genetic information (covered by 'genetic resources'), in extracts that may be obtained from these resources, or in the synthesis of substances that originally have been obtained from the resource.

Element 6: In most cases, collection for commercial purposes will involve an applicant wishing to obtain a patent on the biological material or derived material. Therefore, Element 6 may discourage commercial prospecting. Policy-makers may want to treat applications for different kinds of intellectual property rights differently. For instance, because Plant Breeders' Rights are not as exclusive as patents, there is less reason for parties to obtain permission when applying for Plant Breeders' Rights. Others feel, however, that any proprietary claim over the material should require permission.

Article 9: Conditions for benefit-sharing

Option 0

no provision [i.e., there are no mandatory benefit-sharing conditions to be included in the agreements]

Option 1

Access shall be granted if the following minimum conditions for benefit-sharing are satisfied:

Element 1

flat fee [at a rate to be set by detailed regulation under the authority of this statute]

Element 2

royalties on downstream commercialization [at a rate to be set by detailed regulation under the authority of this statute]

Element 3

recognition as a partner in intellectual property ownership obtained on derivatives of the supplied material

Element 4

access, free or at concessionary rates, to commercial products [at a rate to be set by detailed regulation under the authority of this statute]

Element 5

transfer of technologies [the parameters of which would be identified in detailed regulation under the authority of this statute]

Element 6

training/capacity-building for local partners [see commentary for uses of relevant technologies, general training, etc.]

Element 7

[other benefits to be identified in detailed regulations under the authority of this statute]

Commentary

The main difference between Option 0 and Option 1 is that the latter includes minimum terms and conditions for benefit-sharing to be included in all access agreements forged pursuant to this legislation. Making such conditions mandatory has the advantage of protecting the interests of suppliers (be they individuals, communities, etc.) that might not enjoy bargaining power equal to that of access-seeking parties. On the other hand, if the conditions are unrealistically demanding, making them mandatory may drive potential access-seeking parties away.

Article 10: Revocation of access permit

Option 0
no provision

Option 1
Access permits may be withdrawn by the competent authority:
a) when there is evidence that the collector has violated any of the provisions of this Law;
b) when there is evidence that the collector has failed to comply with the conditions of the access permit; or
c) for reasons of overriding public interest, including protection of the environment and of biological diversity.

Part Four: Application/authorization procedure

Article 11: Application

The applicant shall commence the procedure by delivering an application for access, in accordance with Article 7, 'Application for access permit', to:

Option 1
the competent authority.

Option 2
the competent authority. The competent authority will register the application in a public register and circulate the application to all local communities and individuals concerned.

Option 3
the local community / individual concerned.

Option 4
the local community / individual concerned. The application shall also be sent to the competent authority, which will register it in a public register.

Commentary

The four options listed here correspond to the options in Article 6, 'Parties authorized to grant consent', above.

Article 12: Public consultations

Option 0
no provision

Option 1
The applicant [may] [shall] hold public consultations with the relevant interested parties and governing bodies, including local communities, regarding the access applied for. Public consultations shall be conducted in a transparent manner and shall comply with due-process requirements, including public notice within a reasonable period. The applicant [may] [shall] work in consultation with local community leaders to select a venue for the hearing, to develop an agenda, and to co-chair consultation meetings in those instances where community resources are the subject of the application. Anyone may attend the consultative meetings and all shall be allowed a reasonable opportunity to express their views.

Commentary

It is suggested that countries enact specific requirements on public consultations. As in the case of the consent by indigenous and local communities, these requirements should ensure transparency and compliance with due process. They should include requirements on how public and prior notice is to be given, the nature of the information that should be disclosed, the modes by which such information should be disseminated, the periods of time for public notice to be effective, the languages in which information should be given and consultations undertaken, and the circumstances under which public hearings would be required. The required procedures would have to strike a balance between the public's interest in transparency and the possibility that many access-seeking parties may not want to reveal too much about their work in order to protect their market advantage. Again, too-onerous provisions may drive would-be applicants away.

Article 13: Time limits for decisions

Option 0
no provision [i.e., no time limit]

Option 1

The competent authority shall indicate receipt of the application within [time limit], and shall take a decision within [time limit] of the date of receipt.

Commentary

Option 0 leaves all parties free not to respond to access applications. It may be contrary to the interests of the country not to have such an obligation, inasmuch as that could lead to a national reputation for inefficiency. If individuals or communities alone were to have the authority to consider applications (without the government), then perhaps an obligation to respond would be inappropriate. In such a case, perhaps individuals and communities should be free to ignore an application.

Option 1 requires parties to respond to applications and sets a specific time. It is possible to create a positive obligation for governmental parties to respond, but not to extend that obligation to communities or private persons.

Part Five: Civil and criminal process

Article 14: Civil and criminal process

It would be premature at this point to provide a lengthy analysis of potential offences, defences and remedies that could flow from the options set out in Parts One to Four. To provide a sense of how these very important issues could be addressed in a national access law, we include the following short discussion.

Paragraph 1: Offences; causes of action; intellectual property interventions

Commentary

The access law options set out in Parts One to Four lay the foundations for the definition of numerous civil causes of action and criminal offences. Perhaps the most obvious offence flowing from these provisions would be that of knowingly using biological material that falls within the scope of the legislation (Article 2) for uses that are not exempted (Article 3) without first obtaining the permission of the consent-granting authorities (Articles 6 and 7). Another example, based on the procedural requirements (Article 7), would be the act of wilfully misleading a community with respect to any details included in the access application. Both of these acts could be defined as giving rise to civil or criminal liability.

Many of the provisions in this access law collection could work in tandem with provisions in other collections of legislative options that the Crucible Group has cited in this volume. For example, the legislative options concerning (a) patents for biological innovations, and (b) Plant Breeders' Rights include provisions that prevent parties from obtaining intellectual property rights for things that are derived from biological material not collected in compliance with national access laws. (See Topic 3, Section 2, Articles 12 and 13, and Topic 3, Section 3, Article 19, respectively.)

One very serious limitation on the ability to enforce standards is that the offenders, being foreigners, would be outside of the national legal jurisdiction unless they were actually apprehended conducting wrongful acts on the soil of the implementing country. Consequently, the enforcement of these standards would rely upon international agreements regarding the prosecution of civil suits and extradition.

Paragraph 2: Defences

Commentary

The options set out in Parts One to Four also provide a foundation for defences to charges of improper conduct on the part of access-seekers. For example, in response to a charge of having knowingly obtained material in violation of national access provisions, a defendant could argue that the material in question was exempt from the application of the law by virtue of exemptions such as those set out in Article 3, above.

Paragraph 3: Remedies

Commentary

The range of potential remedies is very wide. In part, the remedy depends upon the body that hears the case. In theory, the remedies could include prohibitions from further collecting, compensation, fines, restitution, probation, jail terms, and/or refusal to grant, or revocation of, patents or plant variety protection.

Paragraph 4: Administrative and judicial fora

Commentary

Where aggrieved parties could seek to have their cases heard would depend in large part upon the remedy they were seeking. For example, if they wanted to prevent the alleged offender from getting intellectual property protection for innovations or plant varieties derived from unauthorized uses of biological resources covered by national access laws, they would make submissions to the patent office or Plant Breeders' Rights office in the jurisdiction where the alleged offender was seeking to obtain intellectual property rights. Of course, the success of this venture would depend upon the nature of the provisions in that jurisdiction's plant variety protection and patent laws. It would require the inclusion of provisions like those set out in Topic 3, Section 2, Articles 12 and 13, and Topic 3, Section 3, Article 19.

Where an aggrieved party wanted an injunction against a party to stop them from engaging in unauthorized collecting, restitution for lost profits, or other civil remedies, they would go to the civil courts. Whether they sought to bring their action in the jurisdiction where the collecting took place, or in the alleged offender's jurisdiction of residence, would depend upon several factors, including reciprocity agreements between the countries concerned, the domestic law of each country, any prior agreements between the parties regarding the choice of jurisdiction in the event of a civil suit, and prevailing principles regarding international conflicts of laws.

If the access law creates criminal or quasi-criminal offences, the aggrieved party could bring the alleged offence to the attention of the state authorities. The state could then decide whether it wished to prosecute the case.

Alternatively, the government could create a specialized tribunal to hear complaints regarding access-related issues.

Part Six: Competent authority

Article 15: Designation of competent authority

Option 0

no provision [i.e., no administrative body is necessary]

Option 1

The administrative duties associated with the obligations set out in this Act shall be undertaken by the [name of ministry].

Option 2

There shall be created a new administrative body referred to in this Act as 'competent authority'. This body shall consist of relevant stakeholders, including:

Element 1

representatives of various relevant ministries

Element 2
representatives of civil society organizations

Element 3
representatives of local and/or indigenous communities

Commentary

The advantage of locating the authority to implement an access law in one ministry is that there would be clear authority and responsibility. However, the very nature of the activity being regulated would require the coordinated action of different agencies such as the ministries of the environment, natural resources, agriculture, health, and the agencies in charge of customs and indigenous peoples' affairs. Civil society and community representation is also important to make the law work and to contribute to its transparency. For this reason, an interministerial, multisectoral mechanism might be most effective.

Topic 2

Options for national laws to protect indigenous and local knowledge regarding biological resources

Contents

Section 1
Introduction

A. Underlying rationale for focusing on indigenous and local knowledge

In the following paragraphs, we set out the four most frequently invoked justifications for making domestic laws to protect, promote and conserve indigenous and local knowledge. Since none of the four is mutually exclusive, they can be read together to form a compelling argument for the development of a range of integrated regulatory policies for the protection of indigenous and local knowledge. Not everyone in the Crucible Group subscribes with equal enthusiasm to each of these justifications. This lack of agreement is not terribly significant, as we are only reporting here, in broad strokes, on the international dialogue that has developed to date on this subject.

A.1 Indigenous and local peoples' human rights/self-determination

In numerous statements and declarations, indigenous peoples (and local community organizations, although to a lesser extent) have stated that they feel that their quest for self-determination is the most important reason to engage in efforts to protect their knowledge. Their collective knowledge is, after all, a critical element of their distinct, self-determined, self-identified cultural existence.

As we will see below, some approaches to protecting indigenous and local knowledge are more supportive than others of indigenous peoples' self-determination. One example of a knowledge-protection strategy that complements the goal of self-determination would be the affirmation of indigenous peoples' customary laws regulating the use and dissemination of their own knowledge and the enforceability of those laws against the greater national community. In this way, while vesting customary forms of protection in the holders of knowledge, indigenous peoples' knowledge governance systems would be recognized and reinforced. Another example would be the formal recognition of the right of indigenous and local communities to determine all natural resource use and conservation strategies on lands that they occupy. This approach would raise the profile of their knowledge of the use of biological resources and ensure its active use.

Some critics charge that political objectives such as self-determination should have no place in the minds of national policy-makers considering options for national laws concerning indigenous and local knowledge.

A.2 Unfair takings

Another widely cited reason to promote, protect and conserve indigenous and local knowledge is that it is being used by non-indigenous and non-local parties in ways that are unfair and unwanted. This position presupposes that existing national laws are inadequate to stop parties from taking, using and reproducing indigenous and local knowledge without first obtaining permission.

Concerns about unfair takings generally fall into two categories. The first category concerns commercially valuable knowledge. Indigenous and local people would like laws that would compel the users of their commercially valuable knowledge to provide them with some form of compensation. The second category concerns sensitive and sacred knowledge. Here the problem is not remuneration for others' use of the knowledge; instead, indigenous and local peoples are more concerned about the long-run secularizing effect that widespread distribution of sacred knowledge would have on their cultures. In this case, indigenous and local people would like to have a law that gives them the power either to prohibit, or to set conditions for, reproduction of this knowledge, whether it has commercial value or not.

A.3 Preventing loss of knowledge

It is important to protect, promote and conserve indigenous and local knowledge because it is disappearing at an accelerating rate. There are many interlinked causes of this disappearance: the eradication of indigenous cultures and populations; the relatively low profile of indigenous and local knowledge versus so-called scientific knowledge; the lost influence of elders in indigenous and local communities; and so on. Policies are needed to raise the profile of indigenous and local knowledge both inside and outside of indigenous and local communities. Such policies would emphasize the importance of the practical use and dissemination of indigenous and local knowledge (quite apart from, or prior to, the issue of who should be able to enjoy the benefits that might accrue from the use of that knowledge).

A.4 Biodiversity conservation

Another widely cited reason for promoting, protecting and conserving indigenous and local knowledge related to biological resources is that it is important for the promotion of global environmental and food security. This position is based on the idea that people who live in proximity to, and are dependent on, local ecosystems for their survival use their natural resources in ways that conserve them (and in the case of genetic resources, in ways that promote genetic diversity and interspecies variation). Consequently, it is in everyone's interest to conserve, promote and protect indigenous and local peoples' knowledge regarding biological resources.

A considerable body of literature and scientific data has accumulated in recent years to support the argument that the natural resource management and stewardship practices of many indigenous and local cultures are more environmentally sustainable than those of so-called mainstream cultures. However, there are exceptions to this rule, and many people are not entirely convinced of the historical accuracy or the prescriptive value of the indigenous/environmentalist paradigm. Some would charge that it is far too simplistic to paint a picture of all indigenous peoples and local communities throughout history and around the globe as environmentalists whose culture and knowledge are more relevant to the sustainable use of biological resources than those of anyone else. These critics charge that the drive to protect indigenous and local knowledge is much more political than it is a logical consequence of universal environmental concerns.

B. Legal sources for the obligation to treat indigenous and local knowledge specially

B.1 International law

The legal obligation to make indigenous and local knowledge the focus of national policy and law-making efforts has many sources. For example, in international environmental law, the Convention on Biological Diversity (CBD) requires signatories to "respect, preserve and maintain knowledge ... of indigenous and local peoples." Pursuant to that requirement, in May 1998, the Fourth Conference of the Parties to the CBD (COP-CBD) decided to create an intersessional *ad hoc* open-ended working group to provide advice to the Parties regarding the "development of legal and other appropriate forms of protection of the knowledge ... of indigenous and local communities". In May 2000, the Fifth Conference of the Parties extended the mandate of this working group and directed it to take steps towards the development of parameters for such legal systems. The Convention to Combat Desertification (CCD) urges countries to "protect, promote and use ... traditional and local technology, knowledge and practices ...". A range of international human rights instruments addresses the question of indigenous and local community knowledge. The Draft Declaration on the Rights of Indigenous Peoples (which is not law yet, but represents the highest expression in an intergovernmental forum of indigenous peoples' aspirations) states that indigenous peoples "are entitled to the recognition of the full ownership, control and protection of their cultural and intellectual property. They have the right to special measures to control, develop, and protect their sciences, technologies and cultural manifestations, including human and other genetic resources, seeds, medicines, knowledge of the properties of fauna and flora, oral traditions, literatures, designs and visual and performing arts." The proposed Inter-American Declaration on the Rights of Indigenous Peoples includes similar obligations regarding indigenous peoples' "right to special measures to control, develop and protect, and full compensation for the use of their sciences and technologies".

Other international legal sources, although they do not mention indigenous and local knowledge explicitly, certainly support the notion that countries are under a growing obligation to introduce policies to deal with indigenous and local knowledge. For example, the International Convention on Social and Cultural Rights (ICESR) includes the right to development and diffusion of science and culture. Pursuant to this covenant, states are obliged to provide measures for the enjoyment of the cultural heritage of indigenous peoples. The International Labour Organisation's Convention 169 Concerning Indigenous and Tribal Peoples in Independent Countries (ILO 169), Article 2, urges signatories to promote "the full realization of social, economic and cultural rights of [indigenous and tribal peoples] with respect to their social and cultural identity, their customs and traditions, and their institutions". Given the dangers of cultural extinction described in *Seeding Solutions,* Volume 1,[8] and the importance of indigenous and local knowledge protection to preserving cultural diversity, the international human rights norm of respect for cultural integrity is offended by the current lack of any such protection. In July 2000, ECOSOC passed CHR Resolution 2000/87 to establish a Permanent Forum on Indigenous Issues. This UN body will coordinate, and assist to unify efforts of, indigenous peoples at the UN to address various issues including cultural, economic, environmental and development rights of indigenous peoples. There is no doubt that the advantages of this new body will be brought to bear on the development of indigenous knowledge protection norms.

B.2 National laws

There are national laws that advocate the protection of indigenous, local or traditional knowledge. For example, the Kenyan Environment Management and Co-ordination Act, 1999, calls for the integration of traditional knowledge together with mainstream scientific knowledge in the context of natural resource management. The Philippines' Indigenous Peoples' Rights Act, in a section entitled 'Right to Indigenous Knowledge Systems and Practices and to Develop Sciences and Technologies', states: "[Indigenous cultural communities and indigenous peoples] are entitled to the recognition of the full ownership and control and protection of their cultural and intellectual rights. They have the right to special measures to control, develop, and protect their sciences, technologies and cultural manifestations, including human and other genetic resources, seeds, including derivatives of these resources, traditional medicines, and health practices, vital medicinal plants, animals and minerals, knowledge systems and practices, knowledge of the properties of fauna and flora, oral traditions, literature, designs and visual and performing arts." (Neither of these national legislative acts provides technical legal clauses that set out the means by which that knowledge could actually be protected.) Thailand's draft plant variety protection legislation would extend intellectual property protection to local farmers' crop plant varieties (which constitute a very specific embodiment of local knowledge).

Several countries either have, or are developing, national access laws that require access-seeking parties to obtain the prior informed consent (PIC) of indigenous and local communities before they can obtain resources, or knowledge associated with resources, located on their lands. While these laws do not create intellectual property protections *per se* for indigenous and local knowledge, they constitute legal efforts to vest exclusive rights of control over knowledge in the hands of indigenous and local communities. There is very little conceptual or justificatory distance between the creation of access laws that include PIC provisions for indigenous and local communities and the creation of intellectual property protections for their benefit.

Finally, there are examples of national laws that implicitly recognize the value of indigenous and local knowledge. The Canadian Environmental Assessment Act recognizes the value of taking traditional knowledge into account when conducting environmental assessment. The European Union is developing policies to preserve particular forms of traditional cultivation. It is relatively predictable and justifiable to progress from recognizing the value of knowledge to recognizing the need to protect it.

B.3 International customary law

Indigenous and local peoples are regularly issuing demands regarding the protection of their knowledge in international policy-making fora. Peoples' declarations such as the Mataatua Declaration on Cultural and Intellectual Property Rights of Indigenous Peoples, the Kari-Oca Declaration, and the Indigenous Peoples' Earth Charter call for national laws to 'control' indigenous and local knowledge and 'protect it against exploitation', and to recognize collective community 'ownership' of indigenous knowledge. Arguably, these declarations, together with the UN conferences, international treaties, and trends in domestic law-making, give rise to an international customary legal obligation to protect indigenous and local knowledge.

B.4 Legal imprecision

The obligations or demands from these various sources are not always the same. The words *respect, preserve, maintain, protect, control, use* and *own* all have different meanings. None of these terms is defined in either the CBD or CCD, nor are they defined in any of the relevant peoples' declarations. Many people working in the field have adopted the term *protect* as representative of this grab-bag of diverse terms. When pressed about what *protect* actually means, however, they usually fall back on the other available terms: *promote, respect, use, own, conserve* and so on.

Nor is there an agreed-upon definition of 'indigenous and local knowledge'. One of the difficulties in trying to define indigenous and local knowledge is that it is a moving target. As knowledge that is constantly changing, it eludes easy capture by legalistic means. Indigenous and local knowledge is not a 'body of knowledge' *per se*. It does not exist as a coherent unified set of information, beliefs and practices that is evenly and generically distributed among all indigenous and local peoples the world over. Instead, it is highly fractured and unevenly spread among peoples, on different continents, in different communities and different groups within communities (e.g., concentrated among women, secret societies and healers.)

Knowledge is subject to competing valuations within and across communities, often with the result that the most highly valued knowledge is that which is held or developed by the more powerful members of those communities. The rate at which knowledge is acquired and transmitted is highly variable and influenced by numerous factors. For example, those members of a community who travel widely participate differently in knowledge systems than those who stay in the community. In some communities, men may herd cattle fifty kilometres away, while women stay in the geographical centre of their community. Similarly, teenagers may leave the community for wage labour and come back with different ideas.

Some of the policies we consider here require a more precise definition of indigenous and local knowledge than others. For example, intellectual property laws for indigenous and local knowledge probably require the greatest degree of definitional precision. This requirement is one of the biggest challenges in attempting to create intellectual property laws for indigenous and local knowledge. In many — some would argue most — situations, the form, content and patterns of use of indigenous and local knowledge is not amenable to being divided and reduced into the kinds of bite-size, acultural pieces that intellectual property laws are best suited to protect. Readers will see that the Crucible Group is constantly struggling throughout this discussion to deal with the fact that the term 'indigenous and local knowledge' refers to a potentially limitless range of beliefs, expertise, information, practices and traditions in as many different forms and content areas, held by an extraordinarily wide array of individuals, specialized groups within communities, communities, peoples and coalitions of peoples. No single legal policy approach can reach and protect all different forms of knowledge that can be described as indigenous and local. Careful attention must be paid to each situation to determine what approach is best suited to different forms of indigenous and local knowledge held by different peoples in different situations.

C. The Crucible Group's main task: analyzing options for national laws and policies

Most of the Crucible Group's work regarding indigenous and local knowledge from this point on is dedicated to identifying and analyzing laws and policies that national policy-makers could implement to advance these goals. We present our work in two sections. The first section consists of a survey wherein we identify a broad range of options and offer a brief description of each. In the second section, we focus on one of these options — *sui generis* intellectual property protections.

Why do we single out intellectual property from all of the options in the survey? For a combination of reasons. First, intellectual property protections for indigenous and local knowledge are very controversial. Since 1992 and the ascension of intellectual property–oriented interpretations of CBD Article 8(j), intellectual property has consistently swamped agendas in community, national and international fora. Second, despite how much has been written on the subject in the last ten years, there has been very little technical legal analysis. Third, many issues that come up in the creation of *sui generis* intellectual property laws for indigenous and local knowledge overlap with issues that arise in the creation of other intellectual property laws that developing countries were supposed to have implemented by 2000, pursuant to the Trade-Related Aspects of Intellectual Property Rights (TRIPs) Agreement. It is important, therefore, that all of these issues be dealt with simultaneously.

We do not mean to suggest that intellectual property laws are potentially a more effective means to promote, respect and conserve indigenous and local

Recommendations

1 In the light of the underlying rationale for focusing on indigenous and local knowledge, governments should recognize that no single policy option could be sufficient to deal comprehensively with the protection, promotion and conservation of such knowledge.

2 Governments should develop an integrated set of policy options based on principles of consultation, representation and coordination. In particular, they should:

a) take stock of existing policies and regulatory bodies that affect indigenous and local knowledge-holders;

b) take stock of the existing customs and practices of indigenous and local communities that affect their knowledge; and

c) consider networking existing relevant regulatory bodies to create an indigenous and local knowledge regulatory network, and the creation of a designated umbrella body to facilitate this work.

knowledge than any of the other options identified in the Section 2 survey. In fact, members of the Crucible Group are divided on the wisdom of governments and communities expending finite resources on the creation of *sui generis* intellectual property rights for indigenous and local knowledge when other policies might be more effective and less complicated. Despite these mixed feelings, however, everyone in the group agreed that it would be a useful exercise to try to develop options for *sui generis* intellectual property laws to treat indigenous and local knowledge. It is our hope that by having engaged in this exercise, a better-defined set of legal norms regarding the treatment of indigenous and local knowledge will eventually be forged out of the wide range of options that are currently open to policy-makers and advocates. The desire to contribute to the transparency of the norm-creating process is what has motivated the Crucible Group.

Section 2

Survey of national policy options to treat indigenous and local knowledge regarding biological resources

Contents

Section 2

Survey of national policy options to treat indigenous and local knowledge regarding biological resources

Introduction

In this section, the Crucible Group provides an overview of policies that national governments could (and sometimes already do) use to promote, preserve, enhance, protect and use indigenous and local knowledge. Despite focusing on national policy-making, we do point out those instances where coordinated international action would make these national options more effective.

Part One: Terms

Indigenous vs. local

Many indigenous peoples are concerned that constant use of the compound term 'indigenous and local' ignores the fact that indigenous peoples have better-defined rights (or at least potential rights) than local communities in many international agreements and national laws. For example, other than in the ambiguous reference to 'local' in the Convention on Biological Diversity (CBD)'s oblique phrase, 'indigenous and local knowledge', 'local' people have not been the subjects of international treaties. The International Labour Organisation's Convention 169 concerning Indigenous and Tribal

Peoples in Independent Countries (ILO 169) speaks of 'tribal' people in addition to indigenous people. Tribal people may be local, or they may not. The point is that, overall, there is comparatively little recognition of the rights and legal existence of local communities compared with indigenous peoples. This is illustrated by the last 17 years of activity in the UN Commission on Human Rights regarding the creation of the Draft Declaration on the Rights of Indigenous Peoples, and the activity at the Organization of American States towards the creation of the proposed Inter-American Declaration on the Rights of Indigenous Peoples. In addition, the body of scholarly commentary on local communities as subjects of international law is relatively small compared with that concerning indigenous peoples.

Part of the reason for the comparative lack of legal attention to 'local' people is that it is difficult to know who can be defined as local. One common explanation for the appearance of the word 'local' in the CBD and the Convention to Combat Desertification (CCD) is that in the cultural history of Africa and Asia, the word 'indigenous' is not very useful — many would argue — to distinguish one segment of society from others. The idea, therefore, is to use the term 'local' to serve constituencies that share many of the same social characteristics and relationships to their country's dominant social orders as do indigenous peoples in New Zealand, Australia and the Americas, but who do not necessarily fit the meaning of that term as it is understood in those countries. But there are other competing, less focused interpretations of who could be local. A local community could be any community that identifies itself as such based on any characteristic its members think is important. At the outer limit of this approach, everyone and anyone can be considered a member of a local community.

The reason why indigenous peoples are concerned with the mixing of 'indigenous and local' is because of the potential that indigenous rights could be undermined or watered down at the national and global level. Indigenous peoples have struggled for recognition of their rights by governments and international bodies for more than three decades. An essential element in this is the recognition of indigenous peoples' right to self-determination, which in international law is accorded to peoples.

The counterargument is that indigenous peoples would lose nothing if policy-makers were to 'piggy-back' local communities' interests on those of indigenous peoples. The exact implications will depend entirely on how a state chooses to define 'local'. A relatively narrow definition would have little or no impact on the rights of indigenous communities, while a broad definition would preclude the possibility of any preferential treatment.

The Crucible Group is refraining (at least in this two-volume set) from making judgements on this issue. Therefore, we will not consider different policies for indigenous peoples and local communities based on this sort of analysis. It is true that the backdrop of international law is more developed where indigenous peoples are concerned. But that does not mean that national policy-makers should not consider these options for 'local' communities as well, or at least make a serious effort to consider various definitions of 'local'.

Indigenous and local knowledge: different levels of specificity required in different approaches to 'protection'

In the collection of *sui generis* intellectual property options in the next section, a great deal turns on the ability to define with precision discrete subsets of indigenous and local knowledge. This is necessary in the context of intellectual property because in most kinds of intellectual property laws (with the exception, for example, of trade secret laws), rights are conferred on the owner of isolable instances of knowledge, and because those rights may be universally asserted against any and all others within the jurisdiction. The same is not true of many other options set out in this section, which require only a general idea of what is meant by the term.

Part Two: Policies

1. Policies to ensure cultural survival: The survival of indigenous and local cultures is a prerequisite for the promotion, protection, use and conservation of indigenous and local knowledge. Self-evident as this may seem, *the biggest single threat to indigenous and local knowledge is the disappearance of indigenous peoples, local communities and their cultures.*[9] Indigenous peoples and local communities maintain that their knowledge is an integral part of their cultural heritage and identity. They say it does not make sense to talk about protecting their knowledge without recognizing their right to practise, protect and develop the past, present and future manifestations of their culture. It follows that the most important strategy that national governments could adopt to protect indigenous and local knowledge would be to work to remove those conditions that threaten the survival of the indigenous and local peoples within their borders, and to establish measures of support for cultural preservation.

This analysis requires policy-makers to understand indigenous and local community knowledge as an embedded aspect of culture, rather than an isolated, asocial phenomenon.

Knowledge and culture are inextricably intertwined. The dynamic and changing nature of indigenous and local knowledge is dependent upon the dynamism and changing nature of indigenous and local cultures. And the reverse is also true: the dynamism of the culture is dependent upon the dynamism of the knowledge. It is for this reason that knowledge protection strategies must be geared towards fostering the survival of indigenous and local cultures.

An alternative approach would be to take the disappearance of indigenous and local peoples and cultures as a given, and develop strategies to conserve their knowledge before entire cultures or key holders of knowledge (e.g., elders) pass away. In short, this approach is one that values *ex situ* preservation of knowledge without concern for the continued viability of the communities that have developed and nurtured it. Many people, indigenous and local people in particular, are offended about this kind of 'salvage ethnology' approach to the subject of indigenous and local knowledge policy-making. The Crucible Group rejects it outright.

Members of the group appreciate that not all indigenous and local cultures are under equal pressure, and that in recent years, many countries have adopted policies that encourage indigenous cultural survival. Nevertheless, the fact remains that around the globe, indigenous cultures and indigenous peoples are disappearing.

Indigenous peoples have repeatedly made the point that in order to survive culturally within the borders of existing states, they require (a) secure tenure to land,[10] (b) the jurisdiction to govern enough of their own collective cultural lives to maintain their distinct cultural identity, and (c) access to credit and social services. It is not within the competence or the mandate of the Crucible Group to enter into an analysis of how much land, how many self-governing powers or what level of services indigenous and local communities require to maintain their distinct cultures. Nor is it within our competence or mandate to analyze the various factors within each country and/or globally that are contributing to the disappearance of indigenous and local cultures. We must, therefore, be content with making a simple point: indigenous and local communities' cultural survival is fundamental to the promotion, protection, use and conservation of their knowledge systems. Minimum standards of land-tenure security, self-governance and social support are co-requisites for their cultural survival.

2. Policies to engage indigenous and local knowledge-holders in biological resource management: The promotion, maintenance, protection and respect for indigenous and local knowledge about biological resources (or any subject, for that matter) depend upon this knowledge actually being used. Otherwise, such knowledge is ignored, undervalued and marginalized. Engaging indigenous and local knowledge-holders in decision-making about the use of biological resources, therefore, is crucial. Raising the profile of the knowledge in this way will engender respect for it and create incentives for indigenous and local peoples to continue developing and using it.

We provide a short list of five different possible national policies to promote such use. Though not exhaustive, this list provides a framework of analysis to assist policy-makers to think about the issue. Options range from the least to the most inclusive and participatory policies possible within the framework of national laws.

a) Governments could award no additional rights to indigenous and local people to engage in environmental/resource management decision-making beyond those allowed to all individuals. Limiting indigenous peoples' and local communities' rights in this way — by not vesting them with any collective resource management rights as communities or peoples — would deny them any participatory expression of their collective existence in the political sphere. (It also may violate international human rights norms.)

   Many countries do have programmes (or constitutional divisions of power) which recognize, to some extent, collective management rights of indigenous and/or local peoples over some territories. This option, therefore, represents something less than already exists in many states.

b) Governments could prescribe that indigenous and local knowledge-holders will be included in a certain percentage of publicly funded research projects having to do with natural resource management. Indigenous and local peoples could be included as sole researchers, joint research partners, research subjects and so on. (For example, sustainable agriculture has been made one of the key topics of research in the European Commission's Quality of Life Programme under the Fifth Framework Programme. Under this programme, European researchers can have developing-country partners. It would be a small, but useful, step to expand the terms of reference of the programme to include the relevance of indigenous and local knowledge to sustainable agriculture.)

c) Governments could prescribe that indigenous and local knowledge-holders be included in co-management committees regulating practices and service delivery on designated lands or in designated communities. These committees could make decisions regarding biological resources, environment, natural resources, land-use planning, fisheries and oceans, agriculture, health or any other sectors. There are already positive and negative examples of co-management schemes between indigenous peoples and local communities living in protected areas and government agencies. While it is certainly beyond the scope of this volume to engage in such analysis, it would be worthwhile to examine patterns of use and proliferation of indigenous knowledge in those programmes.

d) Governments could prescribe that indigenous and local knowledge-holders have the right to participate in law- and policy-making and public research agendas that have an impact anywhere in the country upon biological resources, environment, natural resources, land-use planning, fisheries and oceans, agriculture, health or any other specified sectors. Their participation in these processes could take several different forms, from simply making submissions to voting on the adoption of policy within whatever forum is making a decision.

This sort of policy expands upon the right set out in the immediately preceding option. Here, indigenous and local peoples would be given the right to participate in decision-making and policy-making at the national level, over a broad range of lands and resources beyond those specific lands upon which their communities' interests are engaged. For example, indigenous and local representatives could have permanent positions on the relevant committees and boards of government ministries and academic institutions wherein such policy is decided. They could be given representative positions in government *ad hoc* and standing committees.

e) Governments could transfer the political power to indigenous and local peoples to govern their own affairs with respect to any one or any combination of the sectors listed above where biological resources are concerned.

All of these options could be implemented in a number of different ways. In these five options, the emphasis is not to ensure that indigenous peoples and local communities are treated as equal to the rest of the national population. Indeed, the idea is to treat them differently — to recognize the special collective nature of their culture and knowledge — and to grant exceptional collective rights of participation to them as an expression of that recognition.

3. Research policies: Governments could adopt policies to encourage research regarding indigenous and local knowledge. This research would include investigations into the content and form of indigenous and local knowledge regarding biological resources; the conditions of its occurrence and disappearance; the state of community technologies; the role that knowledge management customarily plays in community life and relationships; how communities deal with the interface and possible synergies between customary knowledge and alternative 'outside' knowledge; and ways in which both indigenous and local, and 'outside' knowledge might be improved or adapted to suit the needs of both local and global communities.

This research could be undertaken by indigenous and local communities on their own or in conjunction with other research partners (e.g., in the public or private sectors). Examples of joint research could be participatory plant breeding, university- and private-sector research agreements regarding medicinal uses of plants and health care delivery. There is a danger in joint research programs that indigenous and local peoples will be taken on board only after the research questions and parameters have been set. One possible long-term strategy to address this phenomenon would be for the government to assist in the identification of, and provide financial assistance to, a broad agenda for such research, and to make it a condition that indigenous and local community members take the lead in joint research sponsored through this program funding. In either case, research policies should include opportunities for communities to obtain necessary credit, technological support and capacity building to undertake the research. Joint research

parties agreed to in advance (e.g., if both parties agreed by contract to submit to the decision of a mutually appointed arbitrator).

Whether a community registry is set up in isolation or in concert with other registries, a set of legally enforceable rules must be established to govern who can get access to the registered knowledge and under what circumstances. Would it be kept confidential and treated as a trade secret, or published, with parties relying on intellectual property laws to protect knowledge-holders' rights? (We consider both of these options in the following section about *sui generis* intellectual property options for indigenous and local knowledge — Topic 2, Section 3). Alternatively, would the knowledge be made available to anyone who was interested in it, without any restrictions? (In this latter case, the overall purpose of the registry would not be to forge intellectual property protections, but rather to support other priorities such as community consciousness-raising.) Communities' priorities regarding these issues could be expressed in community protocols (as discussed above).

Many communities scattered around the world already keep community registries. Government support for community registries could be considered part of the fulfilment of their knowledge-protection responsibilities towards indigenous and local peoples. Of course, communities should not be forced to keep registries if they do not think it would be worthwhile or if they have unresolved issues related to who would get access to the registry.

Some more formal, legislated variant of registries could be integrated into a national *sui generis* intellectual property law to protect indigenous and local knowledge. In the next section, we consider requiring registration of indigenous and local knowledge in a centralized registry as a precondition for obtaining protection (see Topic 2, Section 3, Article 13).

The form of the registry would depend in large part upon the purpose for which it was created. With respect to biological resources, registries could take the form of plant seed depositories, community gardens, or written, tape-recorded or video databases. There could be relatively tight controls over what could be registered (e.g., only knowledge that satisfied the conditions of protection in a *sui generis* intellectual property system), or the registry could accept any knowledge community members decided to submit for registration. (In this latter case, again, the emphasis of the use of the registry would probably be less closely associated with intellectual property protection.)

c) **exchanges between communities:** Local communities can and should continue to exchange seeds as a form of knowledge protection and dissemination. The first ten-thousand years of progress in the history of crop-plant biological diversity depended exclusively upon local farmers exchanging, planting, selecting and replanting seeds. Seed exchanges continue to be a cornerstone of agrobiodiversity and food security.

In South America, Africa and Asia, indigenous peoples and local communities have been holding 'seed fairs', where they come together across borders to exchange seeds. This exchange involves transfer of indigenous and local knowledge inasmuch as the seeds embody the plant-breeding expertise of the farmers, and the exchanges are often accompanied by information about how best to plant, nurture and harvest them. This practice can be replicated and supported by governments and civil society organizations (CSOs) in other parts of the world.

Of course, there may be restrictions on Farmers' Rights to exchange of patented seeds or seeds protected pursuant to many countries' Plant Breeders' Rights laws. These restrictions are discussed in the context of 'farmers' exemptions' in Topic 3, Section 2, Article 16 and Topic 3, Section 3, Article 23.

8. Marketing policies:

a) **labelling:** Various marketing strategies could potentially contribute to the distribution, promotion and use of indigenous and local knowledge. Perhaps the most obvious would be a labelling programme to celebrate the genetic diversity of food products containing ingredients produced by indigenous and local communities. There are already several eco-labelling and social-justice labelling precedents. 'Fair trade' labels, for example, indicate that products have been bought from producer communities at a specified percentage higher than the market rate, with a guarantee not to drop below a minimum price (independently of how low the market might fall).

   Governments could support such initiatives by lending technical and financial assistance to indigenous and local communities engaged in the production of such commodities, and in some cases, to the ultimate distributors of those products (even if they are not indigenous or local themselves). In some cases, the government might need to take a proactive role in creating a supportive policy environment. For example, 'non-GMO' labels have been challenged in some jurisdictions. Governments could work to pre-empt such challenges.

b) **seed certification:** Many countries have strict criteria regarding the marketability of seeds. Governments should amend seed certification laws that are so strict as to prevent the sale and commercial exchange of genetically heterogeneous indigenous plant varieties.

9. Indigenous and local knowledge ombudsman: There could be a national body to hear complaints from indigenous and local peoples regarding policies that negatively affect their ability to preserve, integrate, maintain and protect their knowledge. The success or failure of this body would depend largely on its powers. If it had the right to prosecute under existing laws and to make reports to the legislature, this could be sufficient. At the same time, it may well be worthwhile for indigenous and local communities to establish such bodies on their own initiative, which could accomplish the same objectives through civil actions and petitions.

In the next collection of options for *sui generis* intellectual property protection, the Crucible Group recommends that any countries creating intellectual property rights for indigenous and local knowledge should create a national indigenous and local knowledge ombudsman to help enforce those rights. The proper and efficient execution of policies set out in this section also requires the expert oversight of a competent authority, preferably comprised of a rotating committee including both government and indigenous and local community representatives.

So far, there really is no appropriate international body to undertake this task in the international arena. In Volume 1 of *Seeding Solutions*, the Crucible Group recommends the creation of a special ombudsman's office at the World Trade Organization (WTO), World Intellectual Property Organization (WIPO) and in the UN generally to deal with issues raised by indigenous and local peoples in the context of those organizations' competences.[15]

Recommendations

1 Domestic governments must acknowledge that the disappearance of indigenous and local cultures is the biggest single threat to the protection of indigenous and local knowledge systems. They should, therefore, conduct formal nationwide reviews to identify those policies and practices that undermine the collective cultural survival of the indigenous and local peoples within their borders.

2 Domestic governments must undertake good-faith efforts to obtain the approval of indigenous and local communities before creating policies that are intended to promote, protect, conserve and maintain indigenous and local knowledge.

3 National governments should create advisory committees comprised of indigenous and local and government representatives with a mandate to review government policies and make recommendations about how those policies should be altered to promote, protect, use and conserve indigenous and local knowledge.

4 National governments should support the creation of an international agreement to create a multilateral fund to support promotion, protection, maintenance and respect for indigenous and local knowledge regarding biological resources. In addition, the international community should make a more explicit statement requiring that existing funds place a priority on indigenous and local community-related projects. Such funds could be administered in concert with the fund currently proposed in the International Undertaking on Plant Genetic Resources for Food and Agriculture (IU on PGRFA) to support *in situ* conservation of PGRFA. The fund would, however, extend to biological resources other than those pertinent to food and agriculture (i.e., non-timber forest products, medicinal plants, etc.).

National governments should support the establishment of the Permanent Forum on Indigenous Issues which was endorsed at the UN Economic and Social Council (UN ECOSOC) Substantive Session in July 2000. This body should be encouraged to create working groups to develop policies and norms concerning the promotion, protection, use and conservation of indigenous and local knowledge.

Section 3

Options for *sui generis* intellectual property laws for indigenous and local knowledge regarding biological resources

Contents

Section 3

Options for *sui generis* intellectual property laws for indigenous and local knowledge regarding biological resources

Introduction

All of the options in this section are focused on the creation of intellectual property rights in indigenous and local knowledge. By intellectual property rights, we mean the creation of rights of control based on the recognition of some connection between 'rights-holders' and the knowledge itself. These rights are fundamentally different from contractually created rights because they establish an enforceable relationship between a rights-holder and everyone else in the jurisdiction with regard to the protected knowledge. Contract rights, on the other hand, create enforceable rights only between parties to a contract (although they may, of course, affect the kinds of activities the contractors themselves may engage in with other parties).

Intellectual property rights vest in 'rights-holders'. (It would be possible to use the word 'owners' if it were clear that the reference was to ownership of a right as opposed to the thing itself. To avoid such conceptual difficulties, we will use the term 'rights-holders' in this volume.)

Intellectual property rights do not always have to be the same in character. For instance, the law may vest rights-holders with exclusive power to stop non-indigenous and non-local parties from engaging in a range of different uses (e.g., commercial use, academic use, etc.). Or the law may create non-exclusive rights, such as the right to receive royalties or attribution as the originators of knowledge when other parties choose to use protected knowledge.

The Crucible Group has limited itself to exploring options for intellectual property laws to apply to indigenous and local knowledge related to biological resources. More specifically, the knowledge we primarily focus on is (1) that which indigenous and local peoples know about the properties and uses of biological resources, and (2) innovations based on those resources. By biological resources, we mean anything that is living, including plants, animals, microbes and the ecosystems of which they are a part.

It could be argued that using the term 'knowledge' confuses the issue, because people have so many different things in mind when they use the term. For this reason, some more precise term such as 'innovations' would be preferable. (In fact, this is the terminology used in the Third World Network (TWN)'s model Community Intellectual Rights Act.) At this point, however, the Crucible Group is not in a position to adopt such a term. To do so would be to pre-empt exactly the kind of analysis we want to encourage through the use of our work. For example, in Part Three, 'Conditions of Protection', we set out a range of conditions of protection for policy-makers to think about. Some of those options are wider in scope than what most people understand by the term 'innovations'. At the end of the exercise, it may well be that policy-makers will select those options that could be described as innovations. In such cases, it would be appropriate for them to state that the scope of their laws is 'indigenous and local innovations regarding the uses of biological resources'. They could drop all reference to 'knowledge'. In the meantime, however, because we have given ourselves the task of setting out a range of options that goes beyond 'innovations' or any more precise term, we will continue to use 'knowledge'.

We do not examine any potential applications of intellectual property laws to fields of indigenous and local community knowledge that are not more or less directly related to biological resources. We do not consider the protection of songs, dances, designs or cultural information (including historical, religious or anthropological information) that would not be directly related to biological resources. Nor do we address *all* of the possible means by which intellectual property laws could be deployed to address indigenous peoples' and local communities' interests in biological resources. We do not, for example, examine elements of *sui generis* intellectual property laws that could potentially vest indigenous and local communities with the right to control the use of their images or names in the marketing of products derived from biological resources. Nor do we broach the issue of indigenous and local communities' potential intellectual property interests in the active chemical components and genetic material that make up the bodies of their community members.

While all of these are certainly important issues, we do not attempt to deal with them in this volume for three principal reasons. First, the Crucible

Group's mandate is limited to considering issues related to biological resources. Second, the Crucible Group took as its starting point a relatively literal interpretation of the term 'knowledge', that is to say, "the fact of knowing a thing, state, person, etc.; acquaintance; familiarity gained by experience."[16] While it may be fair to argue that a graphic representation of a community image or the genetic information about a community member's body should be subject to intellectual property rights and laws, it would be a stretch to call those images or genes 'knowledge' *per se*. Third, no single intellectual property law can protect all of the different kinds of intellectual property interests that indigenous and local communities arguably should have in biological resources. Given the varied nature of the subject matter, different kinds of legal mechanisms would be required to define them at law and to vest communities with control over their use. Legal elements considered here to protect communities' knowledge would probably not be well suited to protect communities' intellectual property interests in their images or genetic material taken from their bodies. This work is important, and we encourage others to undertake it. For the time being, however, the Crucible Group is refraining from analyzing legal elements that might be used in pursuit of establishing intellectual property rights over these kinds of subject matters.

That said, it is worth noting that many of the options we consider here regarding knowledge related to biological resources could potentially be extended, at least by way of analogy, to some innovative cultural phenomena such as songs, dances and stories.

From a legal and technical point of view, it is impossible to confer blanket *sui generis* intellectual property rights on all indigenous and local knowledge regarding biological resources *en masse*, as a single, unified body of knowledge. It is far too broad a class of knowledge, with far too many knowledge-holders, to be dealt with in that way. Nor does it actually exist as a unified body of knowledge. Consequently, one of the biggest challenges is figuring out how to define, with legal precision, what aspects of indigenous and local knowledge regarding biological resources can be made the subject of intellectual property laws, and under what circumstances.

Parts Two and Three of this collection, therefore, are dedicated to defining the range of subject matter that could be included within the meaning of indigenous and local knowledge related to biological resources. In Part Two, 'Definitions', we examine a range of possible different meanings for the important terms 'indigenous and local knowledge' and 'biological resources'. In Part Three, 'Conditions for the grant of rights', we examine a range of conditions under which indigenous and local knowledge concerning biological resources could qualify for protection. These conditions assist in identifying and separating out discrete 'units' or 'instances' (for lack of better terms) of knowledge that would qualify for protection. Policy-makers must decide which conditions they want to include in their own laws. Some of the questions policy-makers must ask themselves are as follows:

a) Should the range of knowledge protectable under the law be limited to particular areas of content relating to biological resources? For example, should it apply to knowledge about the spiritual significance of plants or animals in religious practices, the location of biological resources, or the functioning and interdependence of entire ecosystems? Alternatively, should it be confined to narrower categories of content, such as the practical medical uses of plants, or crop improvement through generations of plant selection and replanting?
b) What additional conditions of protection would be appropriate in order to discriminate among claims for knowledge protection within those subtopics? Should the law require that the knowledge must be novel, original or confidential (or some combination of these conditions) to be protected?

The range of options we present in Part Three is not exhaustive, but it sets up a useful framework of analysis for policy-makers thinking about what conditions of protection for indigenous and local knowledge would be most appropriate in their own national laws.

Having defined the conditions for protection of indigenous and local knowledge regarding biological resources, we then move on to consider options for rights that could be conferred on knowledge-holders with respect to that knowledge. In this context, we consider four classes of rights: (1) exclusive use rights, (2) non-exclusive use rights (e.g., mandatory rents, royalties for third-party use), (3) attribution rights (i.e., the right to be recognized as the developer of knowledge), and (4) whatever customary rights might have been conferred on knowledge-holders according to the customary law of the knowledge-holding community. Customary systems may include all the elements under classes 1–3 above. We also consider whether any or all of these rights should be transferable to third parties, and how long the rights should last. In Part Four, 'Rights conferred', we provide some different options and formulas regarding the duration of those rights.

We then move on to consider options regarding administrative requirements to secure protection in a system of *sui generis* intellectual property rights. Perhaps one of the most controversial issues we address in this context is whether knowledge-holders should have to register their knowledge as a condition-precedent for protection. In a viewpoint box, we include the details of disagreements among members of the Crucible Group about registration. We also provide options about what degree of government involvement is appropriate when community knowledge-holders are negotiating licensing agreements for the use of their protected knowledge.

In Part Six, we alert readers to the fact that comprehensive legislation would have to include definitions of civil causes of action and/or criminal offences in order to give the law teeth. We do not have enough time or space here to provide a detailed analysis, and it might be premature to do so in any case. The content of such provisions would be contingent upon the law's scope, conditions for protection, rights conferred, and the administrative structure created to oversee the law's implementation — all of which must be decided beforehand.

Finally, we consider the relationship of the *sui generis* intellectual property provisions included in this collection of options to other laws that may overlap or conflict with them. For example, it is likely that a *sui generis* intellectual property law that includes some of these options would overlap in scope with laws covering plant variety protection or even laws protecting biotechnological innovations, e.g., pharmaceuticals. We consider options for addressing such overlaps in the seventh and final part of this collection. The first such option consists of requiring applicants for patent and plant variety protection to disclose the origin of the biological materials they have relied on while developing their invention or plant variety. The second option is to require such applicants to prove that they have the prior informed consent (PIC) of the knowledge-holders to use the knowledge they have relied on to make their inventions or plant varieties. To be made operational, these elements would have to be included in national patent and plant variety protection laws. We have included them in our collections of options concerning intellectual property protections for plant varieties and biological innovations in Topic 3.[17]

Before proceeding to the options, we must clarify our use of the term *sui generis* in this section. *Sui generis* literally refers to something that is unique or 'of its own kind'. This obviously includes a galaxy of possible applications. This section focuses on intellectual property law and the possibility of creating new laws (or amending existing laws) to extend intellectual property protection to indigenous peoples' and local communities' knowledge. We acknowledge that the term *sui generis* is also used by some advocates to mean laws and policies to protect indigenous and local knowledge that do not include intellectual property protection. We encourage the exploration of such options, which is why we provide examples of such laws and policies in Topic 2, Section 2 of this volume. However, we want to be clear that in this section, our use of the term *sui generis* is to describe intellectual property laws that are especially designed to extend to indigenous and local knowledge.

As we have stated, Crucible Group members are divided over the idea of creating intellectual property rights in indigenous and local knowledge. That said, everyone agreed that it would be a useful contribution to the overall debate to try collectively to work through a set of options for such *sui generis* intellectual property rights. Some thought that by doing so, they would demonstrate the technical impossibility of making such a system work and thereby steer analysis to more fruitful, non-intellectual property–based approaches such as those set out in Topic 2, Section 2's survey of options. Others felt they could demonstrate that such laws could be made to work and could be used by indigenous and local peoples to gain some control over the dissemination of their knowledge.

No one in the Crucible Group sees the creation of intellectual property rights in indigenous and local community knowledge as creating a biodiversity conservation and indigenous and local cultural survival panacea. Even the group's most optimistic champions of *sui generis* intellectual property laws see them only as an incremental advance in support of (a) the conservation, promotion and protection of indigenous and local community knowledge related to biological diversity, and (b) the conservation of biological diversity generally. We do not, as a group, jointly recommend the adoption of any options included here. Nor do we recommend rejecting any of them out of hand. Ultimately, it is up to national policy-makers and advocates to decide what approach they want to take. We hope that these options will assist them to make their decisions.

Part One: Purpose and scope

Article 1: Purpose

The purpose of these provisions, which create *sui generis* intellectual property rights in aspects and instances of indigenous and local knowledge, is:

Element 1
to vest property rights in indigenous and local knowledge in the holders of that knowledge

Element 2
to provide indigenous and local communities with a means to stop unwanted reproduction and dissemination of sensitive and sacred knowledge

Element 3
to equitably distribute the benefits derived from the use of indigenous and local knowledge in academic and commercial research and development

Element 4
to prevent the loss of indigenous and local knowledge

Element 5
to contribute to indigenous peoples' and local communities' self-determination

Element 6
to conserve biological diversity

Element 7
to contribute to food and health security

Commentary

Often, national legislation does not include statements of purpose. When a law does include such statements, they can be used to interpret the meaning of other provisions in the same legislation when that meaning is unclear or subject to debate. This list of possible purposes is meant to focus policy-makers' attention on the overriding objectives they seek to promote through *sui generis* intellectual property laws for indigenous and local knowledge. Since none of the purposes we list here is mutually exclusive, we have denoted all of them as 'elements'.

The first listed purpose — vesting private property rights in knowledge-holders — is the most technically oriented and least ambitious. The purposes set out in Elements 2 through 7 are wider in scope and more ambitious.

Crucible Group members disagree about the extent to which *sui generis* intellectual property laws could achieve any of these purposes. These disagreements can be divided into two categories. The first is disagreements about whether or not *sui generis* intellectual property laws for indigenous and local knowledge are, at a bare minimum, technically feasible. Feasibility depends upon variables including (1) the scope of knowledge the law is intended to cover, (2) the conditions for the protection included in the law, (3) the rights conferred, and (4) whether or not the law is meant to be retroactive. All of these factors must be taken into account when attempting to determine whether the law creates a

Viewpoint box 4: Assuming that *sui generis* intellectual property laws for the protection for indigenous and local knowledge of biological resources are technically feasible, would they serve useful purposes?

Yes.	**No.**
The whole point of intellectual property laws is to vest private property rights in 'owners' of intangibles. If they were technologically feasible, *sui generis* laws for indigenous and local knowledge would do exactly that. Whether or not that is ultimately useful is another question.	

Vesting private rights over knowledge in communities

An intellectual property law could empower a community to prohibit reproduction of sensitive knowledge if (a) the knowledge in question fell within the scope of the law and satisfied the conditions of protection of the law, and (b) the law conferred exclusive rights of control over that knowledge. This power of control would last as long as the law specified.	The scope of these provisions is limited to things that indigenous and local peoples know about biological resources. A great deal of sensitive and sacred knowledge would probably not be included within the scope of these provisions. It might be possible to capture that knowledge in other *sui generis* intellectual property laws with a wider (or different) scope, but not in these.

Stopping reproduction/dissemination of sensitive knowledge

Many intellectual property laws include public disclosure of the protected subject matter as one of the conditions of protection (e.g., written descriptions or deposits of biological material). These laws create a distinction between third parties' knowledge of the existence of an invention or plant variety and their ability to use it for proscribed activities. If a *sui generis* intellectual property law for indigenous and local knowledge worked in this way, knowledge-holders would have to be content that others would be able to learn about the existence and content of knowledge, but would not be able to use it in ways prohibited by the law. This may not satisfy communities' interests in restricting the dissemination of sensitive knowledge.

That said, trade secrets law (which we approach in a *sui generis* way in Article 7, below) has the capacity to deter public disclosures. Of course, knowledge must have the quality of confidence to qualify for protection in the first place.

Enforcing sharing of benefits from commercial use of knowledge

The success of *sui generis* intellectual property laws to enforce benefit-sharing depends upon the scope and conditions for protection in the law, the rights conferred on knowledge-holders and the duration of those rights. Assuming technical feasibility, there is no reason why such laws could not be used to require parties to compensate knowledge-holders when they use knowledge that is protected under the law.

There are situations where parties have attempted to use existing intellectual property laws to prohibit third parties from appropriating indigenous and local knowledge, sometimes with success. (For a discussion of the strengths and weaknesses of existing intellectual property laws as protectors of indigenous and local knowledge in the cases of basmati rice, ayahuasca and quinoa, and turmeric, see *Seeding Solutions,* Volume 1, pp 21–3 and 83–4).

Preventing loss of knowledge

Vesting legally recognized ownership of knowledge in communities through *sui generis* intellectual property rights will raise the profile of that knowledge and encourage respect for it both inside and outside the knowledge-holding

Using a law to make something into property that was previously part of the public domain (according to a state's domestic law) does not suddenly save it, conserve it, or make people respect it or want to use it. One of the most fundamental problems facing communities. This will make the learning and use of such knowledge a more attractive prospect for the younger members of these communities, thus perpetuating its existence.

The possibility of economic return for the use of protected knowledge by third parties acts as a further incentive for community members to respect their knowledge and to continue to engage in practices that use and generate that knowledge.

Indigenous and local knowledge-holders will be more willing to disclose otherwise secret knowledge once they know that *sui generis* laws can give them control over how their knowledge gets used. In this way, intellectual property laws encourage the disclosure, use and proliferation of knowledge that might otherwise be lost.

indigenous and local knowledge-holders is that most of the world still denies the value of that knowledge in the first place. Fencing off their knowledge does nothing to protect it from being eroded, undermined, ignored or at risk of being lost.

To protect knowledge, intellectual property laws must divide it up into bite-size, appropriable pieces. Over the long run, repeated attempts to make knowledge 'fit' intellectual property law criteria would result in an alteration of the nature of the knowledge that communities would generate, thereby contributing to its loss.

Most (possibly all) forms of intellectual property protection cannot protect knowledge while it changes; the duration of protection is exhausted when the knowledge changes from what it was when it was originally protected. New protection may be obtained for changed knowledge, but the protection does not stay with knowledge that is changing. In this way, intellectual property laws 'freeze' knowledge, undermining its dynamism and cultural relevance.

The reasons indigenous and local community knowledge-holders keep their knowledge secret rarely have to do with concerns over its subsequent misuse for commercial purposes. Reluctance to share it is based more on the fact that it is supposed to be restricted to a relatively small number of privileged people in the first place. The kinds of controls intellectual property rights provide will not act as incentives to publish this kind of knowledge.

Promoting self-determination

To the extent that a *sui generis* intellectual property law is structured to reaffirm (a) customary laws regarding the use and dissemination of knowledge, and/or (b) communities' rights to decide for themselves what knowledge should be protected and how, then it could be said to advance self-determination.

To the extent that a *sui generis* intellectual property law could be used to stop others from using knowledge that communities would rather keep to themselves, it provides those communities with a measure of control over their relations with the rest of the national community. Such control is an element of self-determination and collective cultural sovereignty.

It is extremely difficult, if not impossible, to structure a national *sui generis* intellectual property law in such a way that it incorporates indigenous peoples' and local communities' customary laws and practices regarding knowledge. To do so would require the imposition of a different legal standard by each community.

Transfers of jurisdiction to indigenous and local communities to decide for themselves how they will govern knowledge-protection issues inside their own communities (and with the rest of the national community) is certainly an option. It is not, however, something that can be achieved through intellectual property legislation. It is more an issue of constitutional divisions of power.

The Mataatua Declaration on Cultural and Intellectual Property Rights of Indigenous Peoples and the Draft Declaration on the Rights of Indigenous Peoples both include the legal recognition of indigenous communities' intellectual property rights as elements of indigenous peoples' self-determination.

Conserving biological diversity

Because indigenous and local communities are dependent upon local ecosystems and live close to them, they interact with the ecosystem in such a way as to improve species diversity and interspecies variation. *Sui generis* intellectual property laws can enforce the sharing of benefits when downstream researchers use indigenous and local knowledge related to biological resources in their research and development. Consequently, these laws will provide an incentive for those communities to continue to engage in activities associated with the propagation of biological diversity.

Nine years after the Convention on Biological Diversity (CBD) was signed, no one still believes the market-oriented fantasy that remuneration for the use of indigenous and local knowledge will pay for communities to continue living sustainably on the land (and to keep coming up with environmentally sustainable innovations). In fact, it is just as likely that such laws would create incentives for the overexploitation of biological resources associated with protected knowledge (e.g., plants with medicinal properties).

Many indigenous and local communities no longer live close to local ecosystems. Regarding those who do, there is evidence that too-close dependence on local ecosystems can sometimes lead to an overall depletion of resources. This occurs when the demand for a particular local resource becomes greater than the local supply and there is no possibility of introducing a foreign substitute.

practically enforceable set of rights and obligations with respect to a subset of knowledge. If so, the law is technically feasible. If not, then the laws cannot be made to work. This entire section consists of an exploration of technical elements that could be included in a *sui generis* intellectual property law to protect indigenous and local knowledge. We leave conclusions about technical feasibility for readers to make after reading through this section.

The second category of disagreements concerns whether or not *sui generis* intellectual property rights would actually advance any of the purposes set out in Article 1 (assuming that they were technically feasible). The high points of these disagreements are set out in the following viewpoint box.

Article 2: Scope

These provisions apply to indigenous and local knowledge related to biological resources.

Commentary

This article is designed to identify the range of subject matter that a *sui generis* intellectual property law can cover. The scope of the following options is limited to knowledge *related to biological resources*. As stated in the introduction to this section, we have made practical decisions about what we could attempt to cover in this exercise. In principle, there is no reason why intellectual property law could not cover a wider range of subject matter.

It would be possible to use some term other than 'indigenous and local' to describe the knowledge that is protected by these provisions. Although that is the term used in the CBD, the CBD does not require implementing countries to use it. For example, the same provisions could apply to 'tribal', 'customary', 'traditional' or 'community' knowledge.

Similarly, as we have stated, it might also be possible to use some term other than 'knowledge' (e.g., 'innovations'). From a technical point of view, the actual name attributed to the subject matter protected by the law is not terribly important since any of these terms could be defined in any number of different ways. The Crucible Group uses 'indigenous and local' the most familiar and 'knowledge' because it is familiar and sufficiently wide in scope to encompass all of the options we set out in Part Three, 'Conditions for the grant of rights'. In the end, the real trick is figuring out what subject matter, under what circumstances, can be protected, and who will hold property rights in it.

We appreciate, however, that there are political reasons for identifying the knowledge through the use of different terms in different countries. For example, Indonesia and China do not recognize the existence of 'indigenous' peoples within their borders, even though they may very well recognize that there are marginalized, culturally distinct peoples within their territories that should benefit through the implementation of laws such as these. In other parts of the world, such as North and South America and Australia, it would probably make most sense to use the terms 'indigenous' or 'aboriginal', inasmuch as the underlying justification for the protection of that knowledge will be associated with the social, political and legal status of indigenous or aboriginal peoples in those parts of the world.

Part Two: Definitions

Article 3: 'Indigenous' and 'local'

For the purposes of this Act, a reference to 'indigenous' and/or 'local' shall refer to communities and peoples, and the cultures of communities and peoples that:

Paragraph 1: Self-identification

Option 0

no provision [i.e., self-identification is not part of the definition]

Option 1

self-identify as indigenous and/or local

Paragraph 2: Other (outside) criteria

Option 0

no provision [i.e., there are no elements to the definition other than self-identification]

Option 1

and satisfy the following criteria:

Element 1
descend from populations that inhabited a state or area within a state at the time of conquest, or colonization, or the establishment of present state boundaries

Element 2
descend from populations that inhabited a state or area within a state for [number] generations

Element 3
subscribe to specific cultural and economic practices and traditions that are integrally connected to their occupation and customary uses of their territories

Element 4
are governed wholly or partially by their own customs or traditions

Element 5
form non-dominant, socioeconomically marginalized sectors of society

Element 6
are ethnically distinct from the rest of the population of the country within which they reside

Element 7
live within a particular political jurisdiction [e.g., municipalities, townships, band-controlled territories]

Paragraph 3: Explicitly included groups

For greater clarity, in addition to any other communities or peoples who are described above, this Legislation shall apply to the following groups within [the implementing country]:

Option 0

no provision [i.e., no list of names of peoples is included in the law]

Option 1

[the law includes a list of names of specific groups and/or peoples within the country]

Paragraph 4: Membership within groups

Option 1

Determinations regarding whether or not particular individuals are included within a local community or indigenous people will be made by referring to both (a) the customary laws and practices concerning membership in the community or people concerned, and (b) international and domestic human rights laws.

Commentary

In this article, we do not purport to be providing options for the creation of a universal, positivist definition of 'indigenous' (or 'local' for that matter). We appreciate that the historical patterns of settlement and the current socioeconomic status of constituent populations within different states makes the applicability of any such definition extraordinarily problematic. We also appreciate that the political advantages of defining 'indigenous' precisely, with a relatively narrow scope of application, or more loosely, with the ability to include a wider scope of peoples, varies from state to state and region to region. What we provide here is a list of options that set out a general framework to be referred to during domestic policy-making processes. Apart from endorsing the principle of self-identification, the Crucible Group does not jointly recommend (or reject) the inclusion of any one or combination of these elements in domestic laws. We purposefully leave that determination to the parties engaged in law-making activities (including, we hope, indigenous peoples themselves) in any particular state.

The most fundamental distinction among these options is between an approach that gives priority to group self-identification (Paragraph 1) and that which vests the power of definition outside those communities, through reference to fixed criteria about what constitutes 'indigenous' or 'local' (Paragraph 2). Not surprisingly, the history of efforts to develop a definition of indigenous in international legal circles has been animated by the tension between these two approaches. For example, most of the options here have been drawn from four primary documents: (1) the International Labour Organisation's Convention 169 concerning Indigenous and Tribal Peoples in Independent Countries (ILO 169); (2) the Draft Declaration on the Rights of Indigenous Peoples; (3) a UN Economic and Social Council (ECOSOC) study entitled 'The Study of the Problem of Indigenous Populations', 1986 (the Martinez Cobo study); and (4) the draft Inter-American Declaration on the Rights of Indigenous Peoples. Each of these documents is characterized by an unresolved tension between the principle of self-determination and reliance on 'external' criteria for deciding who is indigenous or local.[18]

The definitions of 'indigenous' and 'local' have two important functions within a system to create *sui generis* intellectual property rights in indigenous and local knowledge. First, they assist in the description of the knowledge that is supposed to be protected by the laws. Second, they describe the beneficiaries of those protections (beneficiaries are dealt with in Article 13, 'Entitlement to the rights', below.)

By treating indigenous and local together in this article, we are not suggesting that 'indigenous peoples' and 'local communities' should be defined identically (although they could, if that reflected the reality of the situation in a particular country). We have provided options that we think could be included in either, and will allow readers to select for themselves which options to use to build up a definition of each term separately, if they so choose. For example, it is possible to create a definition of 'indigenous' that requires a people to be able to trace their ancestry back to a point before the arrival of colonists (Paragraph 2, Option 1, Element 1). It would be possible within the same legal framework's definition of 'local' not to require tracing ancestry back so far (Paragraph 2, Option 1, Element 2). On the other hand, one could choose to reject the requirement of ancestral connection for both indigenous and local (Paragraph 2, Option 0).

As was stated in the introduction, no international agreements or national laws define 'indigenous and local knowledge' or 'local'. 'Indigenous' is defined in ILO 169, but the definition is controversial and not universally accepted.

Paragraph 1 (Self-identification): This provision would allow communities to determine for themselves which characteristics should be the basis of their own collective indigenous and/or local self-identities. Indigenous and/or local communities could self-identify on the basis of a wide range of shared criteria, such as sociopolitical groupings, political and/or territorial boundaries, clans, tribes, nations, religion, language, and so on. In the absence of self-identification, presumably national laws would specify a fixed set of unifying criteria, and if a community did not coincidentally self-identify based on those criteria, they would be out of luck.

Presumably 'local' would include a wider range of bases upon which a group could found its self-identity, including things like professions, bases of livelihoods (e.g., fishing people), geographically or politically defined communities, and so on.

The historic emphasis on the concept of self-identification has arisen in the context of indigenous populations' struggle for the right to define themselves as 'peoples'. By considering an option in a national law that would extend the right of self-identification to local communities, the Crucible Group does not want to be interpreted as diluting the cultural importance to indigenous peoples of their right to determine their own separate collective identities as peoples.

Paragraph 2 (Other (outside) criteria): There is a risk in looking to external criteria, inasmuch as they transfer some, and potentially all, of the power of self-identification out of the hands of the people concerned and deliver it, transformed into the power of 'other-identification', to the government.

Paragraph 2, Elements 1 and 2: Ancestry/historical continuity might be less important to the definition of 'local' than to that of 'indigenous'. Depending upon the cultural and political history of the country, it might not be relevant to the identification of 'indigenous' either.

Paragraph 2, Element 3: There is a danger that defining a people in terms of their practices or beliefs will 'freeze' their culture in the process. Strangely, Article 8(j) of the CBD pushes one in the direction of looking at practices as evaluative criteria. It specifies "indigenous and local communities engaged in traditional lifestyles".

Paragraph 2, Element 4: The risk in focusing on how a community is governed is that it will often deprive a people of recognition as indigenous or local because they were previously stripped of their self-governing rights by colonial powers. That would not be a fair basis for denying them the right to be identified as indigenous or local.

Paragraph 2, Element 5: The difficulty of including poverty among the criteria is that it would create incentives for communities to appear poor enough, or marginalized enough, to fit within this meaning of 'indigenous' or 'local'. On the other hand, this criterion directly addresses the motivation that many people have for supporting this sort of law in the first place: to improve the economic livelihoods of poor people. It would have the further quality, if considered in isolation from the other elements listed, of being completely free from ethnic identity. Conferring benefits based on racial or ethnic identity can be fraught with potential ethical and practical problems.

Paragraph 2, Element 6: As stated above, mixing ethnic identity and legal benefits or responsibilities is potentially problematic. On the other hand, it appears to be implicit in a great deal of the discussion of this issue. Whether or not it is eventually included in national legislation, policy-makers and advocates will have to give it active consideration.

Paragraph 2, Element 7: This provision is similar to that used in the Philippines' EO 247 and accompanying regulations to define 'local'. This definition may lead to problems, however, where the 'basic political unit' is not coincident with the boundaries of a self-defined community, for instance in a situation where a municipal boundary divides a self-perceived community into two halves. This problem would be exacerbated in the case of nomadic and pastoral communities.

Paragraph 3 (Explicitly included groups): For greater certainty, the idea here is that national laws might include a list of such groups as convenient shorthand for identifying groups within the country that should be included. Obviously, the use of such terms would depend upon the degree of common understanding that exists within that country as to the meaning of the term. Maximum flexibility would be preserved if the list were left open-ended, with the potential of adding other groups in the future.

Paragraph 4 (Membership within groups): Membership within groups is relevant to resolution of tensions that may arise between individuals and collective rights-holders (i.e., groups, communities, peoples). In some cases, this will be relevant when an individual claims not to be part of a community, and thereby not to be obligated to respect that community's interests in an aspect of knowledge in which he or she claims to have separate and individual rights. In other cases, individuals may want to establish that they are members of a community in order to be entitled to a share of the benefits derived from the community's grant of protection for some aspect of its knowledge. (What that share would be would depend upon the formula for the distribution of benefits adopted by the community as a whole.)

Recommendation

In keeping with the text of the Draft Declaration on the Rights of Indigenous Peoples currently being considered by the UN Commission on Human Rights, we recommend against states importing generic definitions of 'indigenous' from outside the country. Instead, states considering *sui generis* intellectual property protections for indigenous and local knowledge should work with their constituent peoples to determine which groups should benefit from the extension of such laws. In doing so, states should respect the principle that self-identification is fundamental to determining which peoples are indigenous. The same recommendation applies to determining which groups are local.

Depending upon the laws and practices of different peoples, it is possible that people not born into those communities could eventually be accepted into them as members, entitled to a full share of communal benefits, rights and responsibilities.

International and domestic human rights laws would place some outside limits on the grounds upon which collectives could dispossess individuals of membership. For example, it would be contrary to international (and some state's domestic) laws to discriminate against (i.e., exclude from membership) individuals based on sex, political opinion, religious beliefs, marital status, and so on. Balancing individual rights against collective rights in this manner is a politically loaded exercise. We do not purport to have struck any such balance in this paragraph. Our comments here are only meant to alert readers to the fact that striking such balances (or imbalances) is an unavoidable part of creating and administering a *sui generis* intellectual property law for indigenous and local knowledge which vests rights in communities.

Article 4: Indigenous and local knowledge

Option 1

Indigenous and local knowledge is that which is held and/or produced by indigenous and local communities within cultural contexts that can be identified as indigenous and local.

Indigenous and local knowledge can be incrementally developed over generations, by a large number of people, including communities or groups of communities.

Commentary

The terms 'indigenous' and 'local' in the aggregate term 'indigenous and local knowledge' can be interpreted as referring to (1) the identity of the creators/developers/holders of the knowledge, and (2) the cultural environment within which the knowledge has been produced. Most commentary in the field (as well as the text of the CBD, Article 8(j)) appears to assume that they refer to both. The producers of the knowledge must be indigenous or local. The environment or cultural context within which the knowledge is produced must be identifiable as indigenous or local.

Given the definition of indigenous and local knowledge provided here, it might be possible for parties to challenge its protection on the basis that it was not developed as part of a practice that was sufficiently embedded in indigenous and/or local cultures. This could devolve quickly into a very regrettable court debate about what is or is not part of indigenous and local cultures. One way to avoid this scenario would be to define indigenous and local knowledge as that which is held or developed by members of indigenous and local communities, without any reference to the cultural context in which it is generated. While this approach would avoid potentially patronizing determinations about cultural authenticity, it would also drain the definition of its justificatory content. Presumably, at least part of the reason that indigenous and local knowledge is worth protecting is because it is generated in association with, and is expressive of, the cultural ethos of indigenous peoples and local communities. Concentrating on the cultural identity of the knowledge-holders, to the exclusion of any consideration of the cultural context in which the knowledge is generated, strikes some Crucible Group members as inadequate.

Whether or not one is comfortable with consideration of this kind of cultural adjudication, it is an unavoidable aspect of a legal enterprise justified at least in part by a normative connection between cultural context and positive rights and entitlements. One way to minimize the difficulties associated with this exercise would be to structure the law to allow indigenous and local communities to identify which aspects of their own cultures are relevant.

We are not proposing that everything that can be said to fit within this definition can be protected pursuant to *sui generis* intellectual property legislation, or at least the kind of *sui generis* law we are considering here. Instead, the definition provides a general outline of the kind of knowledge that could be subject to such laws. In Part Three, 'Conditions for the grant of rights', we consider several options that national policy-makers can use to precisely identify the indigenous and local knowledge that could qualify for protection.

Article 5: Biological resources

Biological resources include genetic resources, organisms or parts thereof, populations or any other biotic component of ecosystems.

Commentary

The definition comes from the CBD, Article 2.

Part Three: Conditions for the grant of rights

Introduction

As stated above, the success of this enterprise depends upon the law's ability to clearly identify appropriable sub-units, or parts, of the knowledge of indigenous and local peoples. In working towards a set of criteria to identify these parts of indigenous and local knowledge, we pose the following questions:

Should protection be limited to certain subsets of knowledge relating to biological resources?

What, if any, qualitative criteria should be used to discriminate among claims for knowledge protection within the various subject areas? Should it have to have the quality of confidence? Does it have to be embodied in some way? Does it have to be novel? Original? Customarily protected? Should it satisfy some other criterion?

In each of the articles in this part, we follow a similar basic framework. We start with an option (Option 0) that rejects the 'condition of protection' that is the subject of the article. Then we provide an option (Option 1) to adopt some version of that same condition for protection. Generally speaking, our 'Options 1' are as close as possible to the way a condition for protection is used in existing intellectual property laws, but with modifications if they are necessary to make those conditions more suitable for the protection of indigenous and local knowledge. So, for example, under Article 10, 'Novelty', our Option 1 is not a simple reproduction of novelty as it is understood in existing patent law, because as a group we feel that an unmodified novelty criterion simply would not work in the context of intellectual property laws for indigenous and local knowledge. Consequently, our Option 1 for 'novelty' consists of a significant modification of the way novelty appears in patent law (i.e., we refer to novelty *outside the community*).

If we feel that a condition for protection taken 'as is' from existing intellectual property laws is not useful or practicable as applied to indigenous and local knowledge, we do not include it at all. For example, we do not include the condition of non-obviousness, which comes from patent law. There, non-obviousness means 'not obvious to someone skilled in the area of expertise to which the invention relates, at the time the invention was made'. The main reason for the provision is to avoid granting patents for (and thereby restricting the public's right to use) innovations that would naturally come about in any case. To require an instance of indigenous and local knowledge to be non-obvious before it could be protected would be to ignore the fact that much indigenous and local knowledge has been incrementally developed over generations by many people. Even if it were possible,[19] there would be no need to hold indigenous and local knowledge to an obviousness test. Presumably (for reasons discussed in the introduction to this section), there are independent reasons for wanting to create *sui generis* intellectual property protections for indigenous and local knowledge. Presumably, the primary motive for creating such protections is not to encourage indigenous and local peoples to make non-obvious accretions to their knowledge where they would not have done so otherwise. We are probably interested in protecting their knowledge and cultures in a way that respects how they are already being produced.

After identifying the two extreme positions with regard to a condition of protection in Options 0 and 1, we then look at ways to create a middle ground between them. In many of the articles, our approach to creating this middle ground has been to further redefine and modify those terms that are adopted in Option 1 to make them better suited to the unique nature of indigenous and local knowledge. (In such cases, we provide alternative definitions under a second definitions paragraph in the same article.) In other articles, our approach has been to create an entirely new option that cuts across the distinctions between Options 0 and 1. (In these instances, we present the middle-ground options as Options 2, 3, and so on.)

One is not free simply to choose Option 0 in every case. One cannot overcome the problem of having a system that looks too much like existing intellectual property law by simply rejecting each condition for protection that is currently being used in those laws (e.g., novelty in patent law, originality in copyright, etc.). It is necessary to have some effective criteria for identifying the knowledge that will be made the subject of rights established in a *sui generis* intellectual property law. Therefore, if one is going to reject these criteria, it is necessary to have something else to put in their place. Otherwise the system will not work. It will break down before it gets started, because there will be no basis upon which to adjudicate competing claims for what should and should not be protected pursuant to the law.

Furthermore, while reading through this part, remember that we are only addressing the issue of possible conditions for protection, and not the rights conferred upon intellectual property holders who are granted such protection. We have purposefully separated these two issues in order to emphasize the potential for flexibility when linking conditions for protection with different rights.

Article 6: General requirement

Protection shall be granted for indigenous and local knowledge complying with the requirements as set out below.

Article 7: Subcategories of indigenous and local knowledge related to biological resources

Option 0

no provision [i.e., there are no restrictions on the content of knowledge that could be protected other than those that are inherent in the options for a definition of 'indigenous and local knowledge' and 'biological resources' in Articles 3 and 4 above]

Option 1

In order to be covered by these provisions, the indigenous and local knowledge related to biological resources:

Element 1
may include information pertaining to the significance and role of various flora and fauna in spiritual practices

Element 2
may include information pertaining to the location of biological resources on indigenous and local communities' lands

Element 3
may include information regarding the relationship between constituent elements in local ecosystems and land or natural resource use

Element 4
must be relevant to the conservation and sustainable use of biological diversity

Element 5
must be associated with practical uses or applications or properties of biological materials

Element 6
must be associated with the improvements or alterations to biological resources

Commentary

A wide range of indigenous and local knowledge can be said to be related to biological resources: for example, folk songs, spiritual beliefs and practical applications. Option 1, Elements 1–6 provide successively narrower categories of knowledge.

Option 0 is for no further subcategorization of knowledge, thereby extending the provisions to all knowledge related to biological resources, including all of the categories set out in Options 2–6.

Option 1, Elements 1–3 are self-explanatory.

Option 1, Element 4 is taken from CBD Article 8(j). It suffers, however, from extraordinary vagueness; it is hard to know what kind of knowledge is meant to be included within such a description. Including this kind of language in a national law could potentially lead to enormous difficulties.

Option 1, Element 5 was inspired by the TWN's model Community Intellectual Rights Act, which protects "cumulative knowledge of technology of the use, properties, values and processes of any plant variety and any plant or part thereof rendered of any or enhanced use or value". Again, narrowing the scope of protected knowledge in this way makes sense where the idea is to protect indigenous and local knowledge within the context of issues related to genetic resources, biological diversity, food safety and optimization of relevant innovation. This option might be interpreted to include biological information that is related to the medicinal application of uncultivated plants.

Option 1, Element 6 is the narrowest option. It would protect only that knowledge which is embodied in an improved biological resource, for example, a novel plant variety or cross-bred cow. It would not include useful applications of existing biological resources, such as medicinal uses of uncultivated plants.

One tough issue to address in this context is whether or not protection would extend to the genes in indigenous and locally bred plants that are responsible for novel phenotypical expressions (independent of the fact that the indigenous or local breeders would probably not be able to identify the genes that are responsible for the phenotypic expressions). It is important to consider the implications of this. It makes poor sense to impute to indigenous and local peoples knowledge that they do not actually have (in this case, of gene structure, biochemical mechanisms and the like). However, the knowledge they do have — of the properties of certain plants — can lead directly to the isolation and subsequent use of the genetic factors that cause such properties. Such isolation and subsequent use may be routine (in some cases) or highly inventive (in others), but in either case typically will depend directly on the original observation, perhaps combined with conservation of the plant resource. It is not a question of giving rights either to the conservers of the plants, or to the technologists who extract the genes; the contributions of both should be recognized, according to the situation.

Article 8: Confidentiality

Paragraph 1: The confidentiality requirement

Option 0

no provision [i.e., confidentiality is not a condition for protection in the law]

Option 1

In order to qualify for protection under this Act, the knowledge must have the quality of confidence.

Paragraph 2: Confidentiality defined

For the purposes of this Act, knowledge will be considered confidential when:

Element 1

it is secret, that is to say that it is not generally known or readily accessible to persons, outside the group claiming protection, that normally deal with the kind of knowledge in question

Element 2

it has been the subject of reasonable steps by the party seeking protection to keep it secret

Element 3

the party obtaining the information must have known, or ought reasonably to have known, that the information was meant to be kept secret or confidential

Paragraph 3: Confidentiality and third parties

Element 1

Downstream parties who obtain the knowledge in good faith and without knowledge of the fact that it arose from a breach of confidence will be presumed to be parties to the confidential relationship.

Commentary

Confidentiality, at least as understood in existing trade secrets or breach of confidence law, depends upon the existence of a confidential relationship between the knowledge-holding party and the party alleged to have taken the knowledge wrongly. Thus, in order for the knowledge to be protected through confidentiality, it must have been disclosed to the offending party in such a way as to bind that party not to disclose it to anyone else. Consequently, if a third party obtains knowledge which a second party relays in breach of confidence, but the third party does not know about that breach, then the knowledge-holder would not be able to take action against the third party. In this way, confidentiality is different from all of the other conditions of protection included in Part Three. All of the other conditions of protection apply to everyone within the jurisdiction, independent of their relationship to the party claiming an intellectual property right.

Of course, the definition of confidentiality is not written in stone, and one of our objectives is to explore possible adaptations of existing intellectual property doctrines to see if they can be made to accommodate the special nature of indigenous and local knowledge. Policy-makers could consider passing a law wherein honest, unsuspecting third parties who obtain knowledge from confidence-breaking second parties can be stopped from using that knowledge. (This is the option provided in Paragraph 3, Element 1.) In some countries, personal property laws work in exactly this way. For example, if an honest, unsuspecting third party buys a car that has been stolen, title to the car remains with the original owner from whom the car was stolen. In this case, the car would be returned to the first owner. That does not mean that the third party has no remedy at all. The third party could still sue the thief who sold it to him for the lost money.

More importantly, the party originally in possession of the knowledge would be able to enjoin third parties from passing the confidential information to fourth and fifth parties.

Nevertheless, there are inconveniences to be considered in applying the presumption of confidence proposed in Paragraph 3, Element 1. Knowledge is different from an object, such as a car. A car typically belongs to someone. Knowledge typically does not. Is it right that in publishing or communicating knowledge, one should be at risk from claims for breaching a confidential obligation of which one was quite unaware? The exchange of knowledge is good in itself and should be hindered as little as possible. There is also a purely practical difficulty: at what stage (if any) would the presumption of confidence stop? If originally confidential knowledge becomes widely known, is any useful purpose served by classifying further use or dissemination of it as unlawful?

The benefit of trade secrets law, from the perspective of policy-makers creating *sui generis* intellectual property laws to protect indigenous and local knowledge, is that its scope is relatively unlimited. In existing trade secrets law, it can apply to business plans, client lists, formulas, and so on.

In this way, it is much more flexible than patent law, limited in scope to inventions; copyright law, limited to 'works' (which is actually a pretty broad category); plant variety protection law, limited to plant varieties; and so on. The importance of this feature depends in part on the other conditions for protection that policy-makers might simultaneously use in their *sui generis* law. For example, if no other conditions are selected, then confidentiality would, on its own, comprehend a far wider potential array of indigenous and local knowledge regarding biological resources than many of the other conditions would permit, either on their own or in combination.

It is important to understand the differences between confidentiality, novelty and originality. Knowledge about a technology that a community or company may be developing could still be confidential, even if it were not novel. Similarly, the fact that a company possesses some formula, even if the formula is not one it invented or created itself, could still be confidential, although it would not be original. In these ways, confidentiality can be distinguished from both novelty and originality.

There are already cases where trade secret laws have been used to protect indigenous knowledge. There is an Australian case where breach of confidence was invoked to stop an anthropologist from publishing information he obtained in confidence from members of an aboriginal community.[20]

Of course, there are other interesting, relevant aspects of trade secrets law, such as the rights conferred on the holders of trade secrets (e.g., they are generally, but not necessarily, non-exclusive), the duration of those rights (no time limit in principle), and so on. But part of the nature of this exercise, as stated above, is to disaggregate existing intellectual property laws and consider those disaggregated elements, with other new elements, to be able to work toward creating an alternative *sui generis* system. For that reason, we will dispense with consideration of rights conferred until Part Four.

Article 9: Form: the embodiment of protectable knowledge

Option 0

no provision [i.e., the knowledge does not have to be embodied in any way, although it could be]

Option 1

In order to qualify for protection pursuant to this Act, indigenous and local knowledge must be fixed or manifest in embodied form, at least temporarily [e.g., physical application, reaction, writing, tape-recording, plant, microbe, animal, chemical reaction, etc.].

Option 2

In order to qualify for protection pursuant to this Act, indigenous and local knowledge must be demonstrable.

Indigenous and local knowledge is demonstrable if it includes any form of storage, custom, methods or practices in which the indigenous and local knowledge is manifest.

Commentary

The purpose of this article is to make policy-makers aware that they must define which aspects of indigenous and local knowledge are to be protected. To qualify for protection, does the knowledge have to be embodied in some form; can it be entirely intangible, without any physical manifestation; or can the law cut across this distinction?

To begin to address this issue, policy-makers must ask themselves some questions. What knowledge are people referring to when they talk about protecting indigenous and local knowledge? Are they talking about knowledge that is embodied in things (e.g., potatoes adapted to local conditions) and in relationships between things (e.g., eating a certain plant to alleviate diabetes; the observed coincidence of increased ocean salinity with lower birth rates in seals)? Or are they talking about knowledge without technical application, such as ideas, language, arguments or general information about their own cultures?

Intellectual property laws are said to protect intangible property, but in order for the intangible property to qualify for protection, many intellectual property laws require some nexus between the intangible property and its physical expression in inventions, plant varieties, documented works, physical performances, and so on. For example, patent laws protect 'inventions' or 'innovations'. While the definition of 'invention' varies from country to country, it always includes some of the following elements: any new and useful

Viewpoint box 5: Must knowledge be 'embodied' to be protected?

Yes. Only embodied property should be protected.

The idea of restricting protected subject matter to that which is embodied is full of complications. There are many instances where the line is not at all clear. For instance, neither patent nor copyright law protects scientific theorems *per se*. But patent laws will protect the technical applications of the theorem. If the theorem can only be applied in one way, then for all practical purposes, patent law does protect the theorem. By limiting what is protected by insisting on its fixation or physical application, the range of material to be protected will be limited to what is reasonably practicable. It is true that 'embodiment' will disqualify a lot of indigenous and local knowledge that might otherwise be protected. But this is a reasonable restriction to produce a practicable system with limited, definable and enforceable rights. Furthermore, at least as far as indigenous and local knowledge concerning biological and genetic resources is concerned, a lot of that knowledge will be embodied.

No. 'Unembodied' knowledge can also be protected.

Sui generis protection for indigenous and local knowledge is necessary because pre-existing categories of protectable intellectual property created by various areas of intellectual property law (i.e., copyright, patent and plant variety protection law) were inappropriate from the start. In creating new protections, therefore, we should not shy away from throwing out old doctrine. A lot of indigenous and local knowledge is embodied (say, in the form of a particular use of a medicinal plant). But much of it is not. Generally speaking, indigenous and local knowledge is communicated orally, and not 'fixed' in the US copyright-law sense of the word. Similarly, a lot of indigenous knowledge (e.g., cultural, spiritual, historical and legal knowledge) is not embodied in some practical application. Extending intellectual property protections to unembodied knowledge has the effect of 'fencing off' knowledge that would not be protected if embodiment or fixation were required. In cases where this could have serious consequences for the free flow of information, it could be argued that indigenous and local knowledge should be subject to exceptional treatment for the following reasons: (1) indigenous and local knowledge is different in form from other knowledge, and therefore requires the benefit of broad criteria, and (2) indigenous and local peoples have been dispossessed of land and economic and cultural power for so long that additional measures in law are needed to promote their interests. A practical system will require demonstrability, but embodiment is not essential for demonstrability.

art, process, machine, manufacture or composition of matter, or any new and useful improvement in any art, process, machine, manufacture or composition of matter.[21] Furthermore, only those inventions that are 'useful' can be protected. Generally speaking, in patent law 'usefulness' is understood as requiring that the invention could physically be made to work, even if it had not been physically constructed at the time of application. Plant variety protection law protects only those plant varieties that actually exist — not theoretical or hypothetical plants. The same is not true of trade secrets, however. Trade secrets do not have to be embodied in any particular form. Nor is it true of publicity rights, rights to personality, or a variety of evolving 'neighbourhood' rights.

None of this discussion is intended to suggest that *sui generis* intellectual property laws for indigenous and local knowledge must only protect knowledge on one side of this divide or the other. In fact, there is no reason why it could not be fashioned to extend to knowledge in both spheres. But it is definitely an issue that must be addressed. Considerable confusion arises in discussions regarding intellectual property protections for indigenous and local knowledge when people fail to consider this fundamental issue.

One formulation (there could be many) that cuts across the embodied/unembodied divide is that of demonstrability. Demonstrability could be defined using an inclusive form of definition so as to retain its open-ended quality. For example, knowledge that had been reduced to writing or symbols could be included, as could knowledge that was demonstrable through ritual or naming practices.

Article 10: Novelty

Paragraph 1: Novelty requirement

Option 0

no provision [i.e., no novelty requirement]

Option 1

In order to qualify for protection under these provisions, knowledge must be novel *outside the group or community that is claiming the intellectual property right.*

Paragraph 2: Means of public disclosure

Knowledge shall be considered novel if it has not been publicly disclosed by anyone outside the knowledge-holding community or communities:

Element 1

by any means [i.e., it is not known outside the community at all]

Element 2

by way of publication via fixed medium (e.g., television, radio, magazine articles, academic publications, etc.) outside the relevant communities [i.e., any form of disclosure other than being formally 'published' would not disqualify it as novel. For example, it could be known simply by word of mouth and still be novel]

Element 3

by way of a commercial transaction outside the community or communities claiming the intellectual property right [i.e., any other form of disclosure would not disqualify the knowledge as novel. For example, it could be printed in an academic journal and still be novel, as long as it is not made the subject of a commercial transaction]

Paragraph 3: Grace period

Knowledge shall be considered novel if it has been not disclosed in the manner set out in Paragraph 2 for longer than:

Option 1

[number] years.

Option 2

[longer or shorter periods of time] for [specified classes of knowledge].

Commentary

Paragraph 1 (Novelty requirement): It may be most appropriate not to include a novelty requirement. Novelty requires that protectable knowledge cannot exist anywhere else (in prescribed forms established by the legislation). Novelty, in the context of patent law, is one way in which the patent monopoly is constituted. In patent law, an independently created invention that is essentially the same as an earlier invention does not entitle the later inventor to a patent. One reason is that the invention is no longer novel. Copyright law, by contrast, does allow for the possibility of independent double origination, i.e., the fact that two works are substantially the same does not prevent the author of each from claiming copyright, provided the works were generated independently of each other.

The imposition of a novelty requirement makes sense if one believes that a strong monopoly reward is needed as an incentive to bring certain kinds of knowledge into the public domain. The novelty requirement helps to constitute that strong reward, and therefore the incentive. However, in the case of *sui generis* indigenous or local intellectual property, we are not trying to use monopoly rights to remedy a market-failure problem generated by the public-good characteristics of knowledge. Instead, we are trying to protect an indigenous or local innovation system that is already in place and functioning and is not the subject of the usual market failure problems.

The novelty requirement militates against the protection of knowledge that is incrementally derived, over a long period, by a large number of people or communities. Communities' boundaries are porous. Unlike high-tech innovations developed within the confines of a formal sector laboratory over a relatively short period, there is very little to prevent knowledge that is held in common within a community from being disseminated outside the community. The problem is exacerbated by the fact that a great deal of indigenous and local knowledge has been held by communities for a very long time (and has taken a very long time to develop), thereby having more time in which to seep out to the rest of the world.

Furthermore, without a novelty requirement, more than one community could make an intellectual property right claim with respect to the same or very similar knowledge (for example, the mixing and application of plant-based medicines).

Nonetheless, if one did decide to include a novelty requirement, it could be defined so as to make it slightly more accommodating of the special nature of indigenous and local knowledge than the way it is currently defined in most patent laws. For example, in Option 1 we have limited the novelty requirement to novelty *outside the group of people or community making the claim.* Accordingly, knowledge can have existed within the community for a long time and still be novel, as long as it is not widely distributed, commercially transacted or published outside the community.

Conceiving of novelty in this way serves two purposes. First, it respects the fact that indigenous and local knowledge may accumulate slowly, over generations, and exist within a community for a very long time before that community decides it wants to appropriate that knowledge through *sui generis* intellectual property laws. Second, it respects the fact that the entire community may be involved in the incremental development of that knowledge. Since the entire community may be the 'author' of the knowledge, it only makes sense that novelty outside the community or collective author should be the proper test.

Paragraph 2 (Means of public disclosure): In addition, 'novelty outside the community' could be defined in such a way as to make it more or less amenable to indigenous and local knowledge. Paragraph 2 sets out elements that become successively less restrictive in terms of what would be recognized as novel.

Paragraph 2, Element 1 is the most restrictive, stating that knowledge is not novel if it has ever been disclosed outside the community, regardless of whether or not it is widely known.

Paragraph 2, Element 2 is less restrictive, precluding from novelty any knowledge that has been the subject of fixed-medium publication.

Finally, pursuant to **Paragraph 2, Element 3,** all knowledge would be novel except that which had been the subject of a commercial transaction before the date of the claim for intellectual property protection. This option leaves open the possibility that a great deal of indigenous and local knowledge that has already been disseminated around the world would still qualify as novel. (The principle of commercial novelty is included in some countries' patent pipeline systems. It is also embraced in the Trade-Related Aspects of Intellectual Property Rights (TRIPs) Agreement (Article 38.2) in the context of determining when design protection should begin).

The definition of publication should be considered in light of the full range of digital technologies and electronic communications. Would knowledge communicated by way of an Internet posting remain 'unpublished' because the Internet is not a fixed medium? This would be an odd result, given that an Internet posting could potentially reach a far wider audience than an obscure journal article. What about limited-access newsgroups or restricted-access indigenous or local electronic networks? Publication, in the copyright sense, means issue of copies to the public, and many forms of digital communication that do not involve a 'fixed medium' do involve an issuance of copies to the public. The existence of a copy in a computer's random access memory, for instance, has been held to constitute a copy by at least one court. Furthermore, not all television broadcasts involve fixation, and many broadcasts are simultaneously fixed at the moment of dissemination.

It might be of significant benefit to indigenous or local peoples to include a caveat to the effect that if knowledge has been published without the PIC of the community, such publication would not act as a bar to novelty. The big issue with such a caveat, however, is whether it should be retroactive. There would be fewer objections to such a provision if it were to apply only to future publications (unless there is consent). There is, after all, a great deal of ethnobotanical and anthropological research that is likely to be considered 'public domain' even though the conditions under which it was done (in colonial contexts, involving people with limited understanding of the scholarly world, with questionable translations, and so on), would not constitute PIC in today's terms.

Paragraph 3 (Grace period): This paragraph allows for the possibility that indigenous and local knowledge could still be novel, even if it has been disclosed in any of the ways set out in Paragraph 2. The longer the grace period, the more the law would be able to accommodate indigenous and local knowledge that has been disclosed outside the community by any of the means set out in Paragraph 2. Such grace periods are included in US and Canadian patent laws (a) to allow inventors a time-limited opportunity to test the market for their inventions before going to the trouble of patenting them, and

(b) to protect inventors whose work is 'leaked' to the public just as they are preparing to file for patent protection. It has been suggested that one could get around the difficulties posed by a novelty requirement simply by having an indefinite grace period.

Where novelty fails owing to breach of confidence

When knowledge becomes known outside the community, its novelty is undermined and, hence, its qualification for protection (if the law includes novelty as a condition of protection). That said, the knowledge-holders might still have some legal recourse if the knowledge was disseminated outside the community through a breach of confidentiality. Of course, this would depend upon confidentiality being included as a condition for protection (1) pursuant to the same legislation as a parallel or alternative condition for protection, (2) at common law, as it does in many countries already, or (3) in another legislative act. For more on confidentiality, see Article 8, above.

Article 11: Originality

Paragraph 1: The originality requirement

Option 0

no provision [i.e., originality is not included in the law as a condition of protection]

Option 1

In order to qualify for protection under this Act, the knowledge must be original.

Paragraph 2: The definition of originality

For the purposes of this Act, knowledge will be considered original when:

Option 1

it is developed collectively by the group, clan or community claiming the right.

Option 2

it has existed within a community since time immemorial.

Commentary

Many, if not most, people would agree that to be protected, indigenous and local knowledge must actually have been developed by the communities asserting rights of control over it. Conversely, very few people would assert that a community should have a right of control to knowledge that it has learned or borrowed from some other community or source. In fact, although none of the most popular draft model laws currently in circulation actually explicitly state that originality is a requirement, many of them imply it.

Paragraph 2 (The definition of originality), Option 1: According to this paragraph, knowledge is original if someone or some group or community developed the knowledge on its own. Accordingly, if ownership of that knowledge were challenged, the community claiming it would have to establish that it developed the knowledge itself. One salient characteristic of much indigenous and local knowledge, however, is that it has been possessed by a community for so long that the community cannot say when or how it was developed in the first place. The community members can only say that they have had the knowledge as far back as they (or anyone else) can remember, that is to say, since time immemorial. In such cases, it would be impossible for a community to prove that it actually developed the knowledge itself. It is for this reason that we have introduced an alternative definition of 'originality', one that respects the fact that traditional knowledge has often been held since time immemorial. Of course, this alternative definition of originality raised evidentiary problems of its own. For example, how does any group prove that it has always known something? How can a party challenging such a claim prove otherwise?

Unlike novelty, originality would not prohibit more than one community from claiming intellectual property rights over the same instance of indigenous and local knowledge, as long as the claimants all developed the knowledge independently (Paragraph 2, Option 1), or have held the knowledge since time immemorial (Paragraph 2, Option 2).

Article 12: Customary products

Paragraph 1: The customary products requirement

Option 0

no provision

Option 1

To be protected pursuant to this Act, indigenous and local knowledge related to biological resources must be embodied in products that are made or developed in conformance with customary methods and materials.

Paragraph 2: Customary products defined

Option 1

For the purposes of this Act, customary products are those which:

Element 1
are made or developed through the use of materials and technologies that are traditionally used by the indigenous and local people claiming the rights

Element 2
are made within [specify geographic location]

Commentary

Paragraph 1 (The customary products requirement), Option 1 severely narrows the scope of embodiment criteria considered in Article 9, Option 1. According to this article, not only must the knowledge be embodied, it must be embodied in a particular way: in products that are made or developed with traditional materials, in traditional ways, within traditional territories. One spin-off legal question would be: how does one define 'traditional'? If too strictly interpreted, this kind of option would reflect a 'frozen culture/frozen rights' approach to knowledge protection. To avoid this, the provision would have to be implemented with a very loose understanding of what constitutes 'traditional'. Otherwise, it would militate against the principle, recognized in the CBD process and elsewhere, that indigenous and local cultures are dynamic, innovative and adaptive. The effect would be that new and innovative, 'non-traditional' ways of using indigenous and local knowledge to meet biodiversity preservation objectives would be denied protection. One way to avoid the more culturally stultifying applications of this provision would be to allow indigenous peoples or local communities themselves to determine when they are making products in ways that are informed by their own indigenous and local traditions. This discussion echoes that set out in the commentaries to Article 3 regarding the procedural divide between giving indigenous peoples and local communities the power to identify themselves and the salient characteristics of their cultures, and referring that authority to 'outsiders' to make decisions based on objective, fixed criteria.

Another spin-off question would be: what constitutes a product? That would be up to the policy-makers. In the context of biological resources, 'products' could include crop varieties and medicines. In a wider context, 'products' could be art, songs, clothing designs, stories, etc.

Paragraph 2 (Customary products defined), Element 2 is inspired by the law of geographical origins. This form of protection is most commonly associated with the use of names of places of origin of French wines (e.g., Champagne or Bordeaux). Certain minimum legal standards concerning the protection of *noms d'origine* or geographical origins are set out in TRIPs, Section 3, 'Geographical Origins'.

Viewpoint box 6: Is it possible to create new, or adapt existing, conditions for protection (i.e., novelty, originality, traditional products, etc.) to define units of knowledge for intellectual property appropriation?

It won't work.

The evaluative criteria that define what knowledge will be protected under patent, copyright, plant variety protection and design laws exclude all but a small proportion of total knowledge. In this way, they preserve a necessary, utilitarian balance between restricted areas of protected knowledge and the freely manoeuvrable public domain. Throwing off the balance in favour of extending intellectual property protection to a wider proportion of knowledge would freeze up the flow of knowledge overall. Custom-made criteria that would 'catch' the indigenous and local knowledge that otherwise 'falls through the cracks' of patent, copyright and plant variety protection laws would have to be cast so wide that it would ruin the balance. Everything would have to be protected, and the flow of information would grind to a halt. What's more, those criteria would be so loose that they would be unenforceable.

It won't work (another reason).

Once you throw out the established intellectual property categories of copyright, patent and plant variety protection laws, you have to start all over with new categories. It is very unlikely that any one set of evaluative criteria are going to apply to all areas of indigenous and local knowledge that indigenous and local communities would like to protect. Instead, special areas of intellectual property law for different areas of indigenous and local knowledge would have to be developed. But each community has different priorities. That would mean a whole new set of laws for each community. At a certain point, it becomes impossible to cater to the cultural particularities of every different group that generates knowledge in a country.

It shouldn't be done.

Protecting indigenous and local knowledge with intellectual property involves disaggregating indigenous and local knowledge into bite-size pieces so that it can be packaged for commercial evaluation and consumption and legal enforcement. Once you have done to indigenous and local knowledge whatever is necessary to make it fit into the intellectual property mould, it would not be recognizable as indigenous and local knowledge anymore. It would lose its place within the interconnected web of holistic indigenous and local cultures. Despite the best intentions of the people advocating its use, intellectual property ultimately 'colonizes' indigenous and local knowledge.

It will work (at least, give it a try).

While there may never be an intellectual property law panacea for indigenous and local knowledge, there are instances where, using novel evaluative criteria, it is definitely possible to extend worthwhile protections to indigenous and local knowledge. What's more, those criteria could be crafted in such a way as to include aspects of indigenous and local customary law and practices. In this way, *sui generis* intellectual property protections can contribute to a reaffirmation of indigenous and local cultures and practices. While the enforcement of some criteria for protection may seem a little 'out there' at first, there are important legal precedents in many countries for the adjudication of similar issues.

Part Four: Rights conferred

Article 13: Entitlement to the rights

Paragraph 1: Individual/collective rights-holders

Option 1
The rights conferred under these provisions in association with knowledge that satisfies the conditions in Part 3 may vest in:

Element 1
indigenous and local individuals

Element 2
indigenous or local collectives, communities, or groups of communities

Element 3
the members of groups or communities that those communities' own customs would identify as the most appropriate rights-holders of the knowledge in question

Element 4
third parties to whom rights in the knowledge were transferred [or shared] by previous rights-holders

Paragraph 2: Multiple rights-holders in the same knowledge

If two or more [indigenous or local individuals] [peoples or communities] are concurrently entitled to protection in the same knowledge pursuant to this Act, then those parties shall be deemed to hold their rights:

Option 1
jointly.

Option 2
independently of one another.

Paragraph 3: Registration of knowledge

Option 0
no provision [i.e., there is no requirement to register knowledge as a precondition of protection; the rights vest automatically]

Option 1

In addition to satisfying the conditions of protection set out in Part 3, knowledge must be registered in order to be made the subject of rights.

Examiners of the competent authority shall investigate the application for protection. If the knowledge claimed satisfies the criteria set out in Part 4 of this law, the competent authority shall register the knowledge in the indigenous and local knowledge registry and issue the applicant a certificate acknowledging the rights vested in the applicant.

Option 2

Knowledge-holders may choose to register knowledge that satisfies the conditions of protection in Part 4.

Registration is not a prerequisite for protection. It does, however, create a rebuttable presumption that the registered party is a *bona fide* rights-holder.

Commentary

This article tackles three issues that affect knowledge-holders' entitlement to rights conferred pursuant to a *sui generis* law: (1) whether communities can be recognized as owners, (2) responsibilities of owners to one another if the law allows for more than one rights-holder of the same knowledge at the same time, and (3) registration as a prerequisite to ownership.

This article presumes that a party must be indigenous or local in order to be granted a *sui generis* intellectual property right in indigenous and local knowledge. This presumption flows logically from the definition of indigenous and local knowledge set out in Part One, which maintains that a central characteristic of that knowledge is that it is developed by indigenous and local people.

Paragraph 1 (Individual/collective rights-holders): One of the fundamental characteristics of indigenous and local knowledge is that it is collectively developed and collectively held. Most existing intellectual property laws, however, embrace the legal fiction that individuals are the primary location of socially useful innovative activity, and consequently vest protection rights in individual persons (natural and legal). Indigenous and local peoples could take their lead from corporations in the way that they overcome this problem related to intellectual property laws.

Pursuant to US patent law, patents cannot vest in corporations; instead, they must first vest in the inventors. Many inventors named in patents work in the large research and development departments of large corporations. Of course, corporate employers want to own the intellectual property created by their employee researchers. This is justified in principle, as the employees are already compensated for their innovative work through their employment contract with the corporation. To this end, corporations contract with their employees whereby the employees automatically transfer their interests in the intellectual property they created 'on the job' to the company. In this way, the corporation — a species of collective with its (sometimes) thousands of employees, managers, executives, shareholders, etc. — comes to own intellectual property in its own collective name. In many other countries, intellectual property laws contain a presumption that works made in the course of employment belong to the employer. By the time an application is made, the rights have already been vested in the corporation. Applications for patent and trade mark rights in Canada are nearly always in the name of the corporation, as any look at the Canadian Patent Reporter will indicate. This is more or less the scenario envisaged in Paragraph 1, Element 2.

In many countries, indigenous and local communities can and do create corporations for various purposes. If the *sui generis* law allowed rights to be vested only in individuals, these corporations could be used as collective repositories of *sui generis* intellectual property rights. They could do this in one of two ways. First, communities could arrange for their individual members to apply for intellectual property rights in their own name and then transfer those rights to the community-controlled corporation. Elements 1 and 4 together involve this kind of manoeuvring. Element 1 vests rights in individuals. Element 4 allows those rights to be transferred to third parties.

Some commentators have urged that it would be beneficial, in this scenario, to legally characterize the corporation as holding the knowledge 'in trust' for the community.

Another possibility with a similar sort of result (inasmuch as it has the effect of transferring beneficial ownership to the community) would be for the initial individual grantee of an intellectual property right to volunteer to hold it in trust for the community.

Paragraph 1, Element 2, on the other hand, allows for protection rights to vest directly in the collective body of the community or group of communities. For practical reasons, it would be convenient to vest the rights in some representative body on behalf of the community, or at least to have some body within the community designated to represent the communities' interests in the knowledge to outsiders.

One possibility is that the law could require (or permit) communities to register themselves as communities in order to be entitled to 'own' collective rights in indigenous and local knowledge. The idea of registering communities is found in the TWN's Community Intellectual Rights Model Law, Section 4, 'Registration of local community'.

Paragraph 1, Element 3 cuts across Elements 1 and 2 by deferring the decision about who should be the rights-holder to community customary laws. It may be that customary laws would dictate that a kind of knowledge should vest in the entire community collectively, or it may dictate that only certain individuals within a community should hold the knowledge. For example, it might be customary within the community for specialized knowledge of a certain sort to be held (and controlled by) a secret society, or healers, or women, or hunters, etc. By deferring to local customs in this way, policy-makers would respect the fact that new property laws may disrupt pre-existing community ownership and distribution patterns. In theory, modelling the distribution of new *sui generis* intellectual property rights on pre-existing community customary relationships of responsibilities and rights would minimize this disruption. Of course, approaching the issue in this way would raise a host of related problems. Many communities will not have customs directly relevant to knowledge exchanges with 'outsiders', since the 'market' for this knowledge is relatively new. Furthermore, this approach would require consideration of each community's customary law in each case. This latter issue has the potential to raise transaction costs to an impractical height.

The same law could include the possibility of conferring the protection rights on both individuals and communities, depending on what would be most appropriate in different situations.

Paragraph 1, Element 4 creates the possibility that communities that developed the knowledge and originally obtained *sui generis* intellectual property protection for it can transfer their rights in that knowledge to third parties. Most commentary on indigenous and local knowledge protection laws rejects this possibility outright on the basis that it violates the integral connection between indigenous and local communities and their knowledge. While this may be so, some counter that this objection is based on a romanticized view of indigenous and local peoples, and that they should be free to alienate their knowledge to third parties when and if they want to.

Paragraph 2 (Multiple rights-holders in the same knowledge): Whether or not a *sui generis* law would recognize more than one party at a time as having rights over knowledge depends upon the conditions for protection included in that law.

For example, if novelty were a condition for protection, then only one party would be entitled to protection over the knowledge, and therefore this article would be unnecessary. On the other hand, if originality were a condition for protection, then there could be several recognized owners. In such cases, it would be necessary to establish what the owners' obligations to each other would be. That is the purpose of this article.

There are basically two choices:

1 The parties can hold their rights with respect to the knowledge jointly, and therefore be obliged to seek the permission of all joint owners before making deals with respect to that knowledge. Benefits derived from deals would have to be shared among all joint owners as well.
2 The parties can hold their rights separately from one another and not have any obligations to one another. Joint ownership is complicated because

there may be several communities who are entitled to protection of certain knowledge. If just one community opposes a proposed deal with respect to that knowledge, all of the others would be prohibited from pursuing that possibility. Separate ownership, on the other hand, suffers from the fact that a single owner could undermine all other owners by making deals without other owners' consent.

Paragraph 3 (Registration of knowledge), Option 0 makes rights vest in knowledge-holders automatically upon that knowledge coming into existence (as long as it satisfied the conditions of protection set out in Part Three). Copyright law works like this in most countries.

Paragraph 3, Option 1 requires parties to have their knowledge examined, accepted and entered into a statutorily created registry before they could be vested with rights concerning that knowledge. We are not referring here to unregulated community registries. For a registry to be used as a mechanism in a national *sui generis* intellectual property law, it would have to be created by statute and fully integrated into the centralized administration of the protection system. Requiring registration in this way would not necessarily have any bearing on (a) the conditions of protection set out in Part Three, or (b) whether or not there could be more than one owner (or one joint owner) of the same knowledge. These two issues are separate from that of registration.

Paragraph 3, Option 2 does not make registration a precondition of protection. It does, however, allow owners to register their knowledge in order to take advantage of a legal presumption, upon registration, that the owner is in fact qualified to be an owner.

Viewpoint box 7: Should knowledge-holders be required to register indigenous or local knowledge as a precondition of protection?

One of the longest standing debates in the context of *sui generis* intellectual property protection concerns the use of knowledge registries. Registries can be used for a number of purposes:

1. they can be used as a mechanism to be included in local resource-use assessments, community awareness-building exercises, and other locally determined priorities that have nothing to do with establishing or defeating intellectual property claims.
2. registries can be used as a means to 'publish' the existence of knowledge and thereby defeat third parties' claims for intellectual property on innovations, plant varieties, etc., on the basis that they are not 'novel' or 'distinct'.
3. they can be used as part of a system to positively assert intellectual property rights over indigenous and local knowledge. In such cases, the use of registries would have to be harmonized with other aspects of the domestic *sui generis* law for the protection of indigenous and local knowledge.

For example, in the type of sui generis system under consideration in this section, it would make sense to require applicants to satisfy the 'Conditions for the grant of rights' in Part Three as a precondition for registration.

The following 'pros and cons' discussion concerns the third listed use of registries as a means to positively assert intellectual property rights over indigenous and local knowledge within a *sui generis* intellectual property system for indigenous and local knowledge.

Yes. Registration would create a predictable business environment.

The chief advantage, from industry's point of view, of a knowledge registry is that it creates a predictable business environment. If something is in the registry, it is protected, and the industrialist knows that he or she will have to seek an arrangement with the registered owner to get access to or use of that knowledge. If the knowledge is not registered, it is not protected, and after a proper registry search, the industrialist will be able to use the knowledge with confidence that there will be no surprise lawsuit halfway through a project.

Yes. Registration would make knowledge easier to protect.

For communities, once their knowledge was registered, they would have a much easier time proving a case of violation. Once it was established that an outside party had relied upon registered indigenous and local knowledge, there would be an automatic legal presumption of violation, and the outside party would have to rebut. It would not be necessary to establish that the user used the knowledge consciously. The existence of the registry would create an obligation on the user to search in advance to ascertain that he or she was not relying on registered knowledge (in the same way that the onus is on persons seeking trade mark protection to first establish that the trade mark name is not already being used by another party).

This is very different from a situation of alleged violation where there is no registry. First, the community would have to prove that the knowledge the user relied upon was community knowledge. Second, in the absence of a registry, which from the user's perspective is an institutionalized searching system, it may not be fair to hold a user liable who did not actually know that he or she was utilizing indigenous or local knowledge. Consequently, the community may also be in the position of having to establish that the user actually had access to, and dishonestly relied upon, their knowledge. (This is what plaintiffs must establish in cases of alleged copyright violations). These comments are premised on the idea that the registry would be national and government-approved and -supported.

No. A community could never register and protect all its knowledge.

One major disadvantage of the registry system is that it places the onus on local communities to register their knowledge. For several reasons, this creates insuperable problems. First, it is very hard for anyone, let alone a community collectively, to determine (and list) what it knows. (Try it.) Second, there is a large potential disjuncture between what a community may think valuable enough to register, and what aspects of indigenous and local knowledge a commercial industry might find valuable. The community could be caught out, realizing only after an outside party has relied on some aspect of their knowledge that it was something they should have registered. Third, registration will require massive organization and mobilization of communities in order to be effective. Unless the new registry system is accompanied by a large infusion of financial resources to support this mass education and mobilization of communities, it is unlikely that many communities will be brought into the registration 'loop' effectively.

In the end, creating a system of registration that communities would not use puts them in a worse position than that of having no registry at all. If industry can use what is not registered, and communities fail to register their knowledge, downstream users are provided with a perfect defence for using indigenous or local knowledge without any recognition that they are doing so.

A registry could be public or confidential.

The comments in the first three boxes are based on the idea that registries of protected indigenous and local knowledge would be maintained like open, publicly accessible patent databases. Another approach would be to keep the registry closed, or confidential. In this case, the registry could be a means of establishing, in accordance with Article 8, 'Confidentiality', above, that the community has in fact identified aspects of their knowledge that they want to keep confidential and taken steps to keep it that way. In this way, the registry would be serving an evidentiary purpose.

If there are to be registries, indigenous and local communities must have 'ownership' in the process.

In a situation where graft and corruption is rampant, false registration would be commonplace, with the potential to divide indigenous peoples among themselves. This has been the unfortunate experience when indigenous peoples were required to register their ancestral land claims in the Philippines as part of a system to establish territorial rights. Some people, more enterprising and scheming than others, fraudulently registered as their own the ancestral lands of others. To minimize the possibility of fraud and bribery and to ensure that the integrity of indigenous knowledge and practices is respected, resources would need to be dedicated to a massive awareness-raising campaign among indigenous peoples and local communities, and the registration should be a mass activity. This should include indigenous peoples and local community members discussing among themselves what should be registered. Inclusion of indigenous and local people in the administration of the registry is also essential. By actively involving the community in the registration process, they will own the process and be more vigilant in safeguarding their rights.

Article 14: The rights conferred

Paragraph 1: Exclusive and/or non-exclusive use rights

Option 1

The owners of knowledge that is protected pursuant to these provisions shall have the right to refuse permission to third parties to disclose, use or sell that knowledge.

Option 2

When third parties use or sell indigenous and local knowledge protected pursuant to these provisions, the owners of that knowledge have the right to remuneration in a form and amount set by regulation [e.g., flat rate, royalties on profits or gross sales, etc.] or negotiated between the parties.

Paragraph 2: Integrity and attribution rights

Option 1

When third parties disclose, use or sell indigenous and local knowledge protected under these provisions, the owners of that knowledge have the right:

Element 1

to be fully acknowledged as the source of the knowledge

Element 2

to the integrity of their knowledge

Element 3

not to be falsely portrayed as developers of the knowledge, and not to have it falsely implied that they endorse the commercialization of a product

Paragraph 3: Customary rights

Option 0

no provision [i.e., no customary rights]

Option 1

The owners of knowledge pursuant to these provisions shall have the same rights over their knowledge in relation to everyone in [the implementing country] as they do within and among their own communities according to their own customary laws and practices.

Paragraph 4: Transferable rights

Option 0

no provision [i.e., the rights in Paragraphs 1–3 are inalienable and cannot be transferred]

Option 1

Indigenous and local knowledge rights-holders may transfer any combination of the rights set out in Paragraphs 1–3 to a third party.

Commentary

This article provides options for the rights-holders over knowledge that is protected by these provisions. The law can be structured in such a way that some of these rights could apply to some kinds of knowledge and not to others. For example, the law might provide a community with exclusive rights with respect to the use of a medicinal plant, but only non-exclusive rights with respect to the use of a plant variety, or vice versa.

This article sets out general 'blanket' rights. In Article 15, below, we list exemptions from these blanket provisions. For example, one option in this article provides owners with the power to prohibit any use of protected knowledge. In the next article, we give options for exemptions from this right, for example, that natural persons may use the knowledge for their own private, non-commercial purposes. Without these exemptions, the rights set out in this article would be stronger than those provided in existing patent laws.

Paragraph 1 (Exclusive and/or non-exclusive use rights), Option 1 confers exclusive rights with regard to the use and reproduction of the protected knowledge.

Paragraph 1, Option 2: Non-exclusive rights do not allow rights-holders to actually prevent use of protected knowledge. Instead, a restrictive condition is created, to the benefit of the rights-holder, that when third parties use the knowledge, they must confer benefits on the rights-holder.

Paragraph 2 (Integrity and attribution rights), Element 1 gives the creator or developer of the protected knowledge the right always to be named as the creator of that knowledge.

Paragraph 2, Element 2 gives the original rights-holder the right not to have its knowledge debased, (or, possibly, not to be altered in any way, depending upon how strongly the right is interpreted). Theoretically, this right would continue even if the knowledge were transferred to a third party.

Paragraph 2, Element 3 gives rights-holders protection against commercial exploitation of indigenous and local communities by marketing professionals who might misattribute the origins of products to indigenous or local peoples for advertising purposes.

Paragraph 3 (Customary rights): The idea here is to extend the application of customary laws relating to the control and use of biological resource–related knowledge within and among communities to everyone in the implementing country within which the law is passed. Then, persons seeking to use knowledge that satisfies the conditions of protection set out in Part Three would have to subscribe to procedures or restrictions that the knowledge-holders would customarily have in place. This option has the advantage of being informed as much as possible by pre-existing priorities in indigenous and local communities. It has the disadvantage of having so many different legal sources (each community presumably has its own rules about what knowledge should be protected and how) that it might be impracticable.

The relationship of conditions of protection in Part Three to rights conferred in Part Four

Certainly the conditions of protection included in a national *sui generis* law have an impact on the rights that can or should be conferred, and vice versa. For example, it would be impossible to confer rights against everyone in the jurisdiction if the sole criterion for protection were confidentiality. As stated above, confidentiality is a condition that creates an obligation on the confidee only (and possibly, through a novel legal amendment, on downstream parties who hear the information in breach of the confidee's obligation). Rights that the confider enjoys with respect to the confidee, however, may be exclusive, non-exclusive and/or attribution and integrity rights.

If originality is the only condition for protection, then policy-makers might want to consider non-exclusive rights and integrity and attribution rights (rather than exclusive rights). This is because almost anything can be original, no matter how banal or obvious it might be. This is part of the rationale for why copyright protection extends only to the expression of ideas and not to the ideas themselves: with originality so easy to achieve from a legal technical standpoint, the world's ideas would all be under private control by now if originality justified this protection. Instead, copyright law cleaves ideas and expressions in two and protects only the latter, leaving the ideas themselves in the public domain. We do not, however, want to be understood as suggesting that *sui generis* intellectual property laws for indigenous and local knowledge should, therefore, import the idea/expression dichotomy. It is clear (at least in the context of knowledge related to biological resources) that what is more important to indigenous and local people is the protection of their ideas and substantive knowledge, and not just the expression of that knowledge. That is why we suggest non-exclusive and attribution rights for the ideas themselves (to engage copyright vocabulary), instead of exclusive rights for just one part of the knowledge (i.e., the expression thereof). Of course, for sacred and sensitive knowledge this approach may not offer a satisfactory degree of protection. For that knowledge, exclusive rights may be the only acceptable form of protection.

Alternatively, when novelty outside the community is included as a condition for protection, policy-makers might be more comfortable considering exclusive rights in association with that knowledge. The fact that something is novel outside the community implies that non-community members would not be deprived of activities that they would otherwise engage in. In this way, novelty represents a somewhat higher threshold than originality.

With respect to knowledge embodied in traditional products, or made by indigenous and local people in a certain region, it is harder to say. The requirement of traditional manufacturing methods and materials makes it seem at first that the products might be so peculiar to the people claiming the rights that it would not be much of an imposition on the rest of the world to confer exclusive rights over their use, reproduction and sale. On the other hand, some products might be simple and already under production by non-community members who developed identical practices on their own. It would not seem fair to prohibit them from engaging in that activity, or to require them to pay a rent or royalty. Out of respect for this kind of criticism, existing geographical nominations law confers exclusive rights with respect to the name of a product only, and not to its actual reproduction or sale under other names by other parties. Would the same protection (for the name of a product alone) suffice in the context of *sui generis* intellectual property protections for indigenous and local knowledge? Or would the protection have to extend to the entire product to be useful? This will depend on the product in question, and on the policy goals that the law is meant to advance. Policy-makers would also have to consider the degree to which the product had already been independently produced within the jurisdiction.

Some commentators in the field have stated that rights in a *sui generis* intellectual property system for indigenous and local knowledge should be non-exclusive (e.g., the community would have the right to be named as the originators of a ritual and medicine, and collect royalties from the sale of a book in which the description is included). Many indigenous and local community groups, however, are adamant that they are much less interested in recouping use-rents for their knowledge than they are in being able to prohibit entirely 'takings' that violate the integrity of their knowledge and undermine its cultural value.

There may be means to address this impasse within a national *sui generis* intellectual property law. The law could confer different rights on knowledge that satisfied different conditions for protection. For example, if knowledge was original and was an embodiment of a traditional product, it could be made the subject of non-exclusive and attribution rights. If, in addition, it was novel outside the community, or confidential, it could be made the subject of some form of exclusive right.

Consider the following example. Designated members of Community A engage in a ritual involving the blessing and application of a mixture of medicinal plants. Researcher X wants to publish a description of that ritual and the medicine. Assume that the ritual and mixture originated in the community, and are embodiments of traditional methods and products. As such, the community would enjoy non-exclusive rights with respect to that knowledge (i.e., the community would have to be recognized as the source of the knowledge and would receive royalties on the sale of the book containing the descriptions). That might be satisfactory. But assume, in addition, that the ritual and medicinal mixture are sacred, and the overriding interest of the community is to prohibit reproduction (i.e., publication) altogether. If the community can establish that the knowledge is novel outside the community, or has been revealed only as a result of a breach of confidence, then the community could be allowed the exclusive right to prohibit Researcher X from publishing (if novelty or confidentiality are part of a graduated scheme of

Viewpoint box 8: How strong should controls over intellectual property–protected indigenous and local knowledge be?

The strength of the rights should be inversely proportionate to the scope of coverage.

The strength of the rights conferred in association with indigenous and local knowledge protection should be related to the breadth of the classes or parts of knowledge to be protected. If the law were to extend to all indigenous and local knowledge, it would make sense to limit the associated rights, because if all indigenous and local knowledge were subject to very strong rights, the overall speed and ease of knowledge transference would be drastically reduced. Consequently, if all indigenous and local knowledge were to be protected, one might decide to apply only the right of attribution. Conversely, for narrower categories of knowledge (e.g., knowledge embodied in specific plants developed by indigenous and local communities), one might grant more restrictive rights, such as the right to forbid reproduction for commercial purposes.

Rights are rights, and should not be undermined by utilitarian considerations.

Indigenous and local peoples have a right to have their knowledge protected, not because protecting it will create incentives for more innovations, but because it is theirs. It does not matter that one potential result of creating strong rights for a wide scope of knowledge would be to tie up the system (if communities steadfastly choose to refuse consent). There is already evidence that existing intellectual property rights are having a negative effect on technology transfers for R & D in the health and agricultural sciences. Why should we trust formal sector innovators with strong rights, and not indigenous and local communities?

conditions for protection within the law, or part of an entirely separate law). A similar result might be obtained if the law included the possibility of an *ad hoc* reference to the customary laws of the community in question to establish whether the community law prohibited reproduction of the relevant knowledge.

This discussion is not intended to be a series of recommendations regarding the relationship between conditions of protection and rights conferred. Its purpose is only to illustrate the range of factors policy-makers must consider.

Paragraph 4 (Transferable rights): This option creates the possibility that a community might sell or give away its *sui generis* intellectual property rights in protected knowledge. A community would also have the option to alienate some rights and retain others associated with the same protected knowledge. Most commentary on the subject assumes that property rights vested in indigenous and local communities concerning their knowledge should not be alienable. To have it otherwise would be to undermine the cultural value of creating the protections in the first place.

In all countries' intellectual property laws that we are aware of, patent rights and Plant Breeders' Rights can be transferred from one party to another. Inventors and plant breeders can sell their rights to anyone they want. Or, when they die, their rights can be passed on through inheritance. After such a transfer has taken place, the second party has all the rights of the original inventor, and the original inventor has none. It is then up to the second owner to determine whether to enforce its rights in the intellectual property and to entertain applications for licences to use the protected invention or plant variety.

The only exception to this general rule of alienability is a variable bundle of rights, known as 'moral rights', found in some countries' copyright legislation.

Generally, national copyright laws including moral rights contain some combination of the following: the right of attribution (to always be recognized as the author of the work); the right to integrity of the work (i.e., even after it is sold, the work cannot be changed in ways that would damage the reputation of the artist); and the right not to have works falsely attributed to them, whether or not the works are copyright protected. Article 6 of the Berne Convention for the Protection of Literary and Artistic Works sets out the first two of these rights.

In France, the law does not allow an author to waive these rights. In Canada, the law allows them to be waived at the will of the author. In the United States, they only exist with respect to a circumscribed number of works pursuant to the *sui generis* Visual Artists Act.

In copyright law, rights not called moral rights are called economic rights. We felt constrained from adopting such terminology here (i.e., 'economic rights'), because of the very different nature of indigenous and local knowledge. A government creating *sui generis* intellectual property laws for indigenous and local knowledge could decide to treat all of the rights conferred on indigenous and local knowledge-holders as moral rights. This is the position urged by the TWN in their Community Intellectual Rights Act, wherein it is stated that all indigenous and local knowledge is inalienable. After all, realizing the commercial potential of indigenous and local knowledge is only one among several reasons to create laws to promote, protect and conserve that knowledge.

The transfer of rights of ownership must not be confused with the creation of licensing agreements that allow licensees to use protected knowledge. In licensing, intellectual property owners continue to retain their underlying property rights.

Article 15: Exemptions from the rights conferred

Paragraph 1: Private use

Option 0

no provision [i.e., private uses are not exempt from the range of uses rights-holders can prohibit]

Option 1

The rights conferred with respect to indigenous and local knowledge in

Article 14 are not infringed by the acts of natural persons [i.e., non-corporations] acting privately and without commercial purposes.

Paragraph 2: Educational research

Option 0
no provision [i.e., using knowledge for purely educational and academic research without first obtaining permission is forbidden]

Option 1
The rights conferred with respect to indigenous and local knowledge in Article 14 are not infringed by the use of knowledge in an educational institution for the purposes of:

Element 1
teaching

Element 2
research [that is not connected to the commercial exploitation of that knowledge]

Paragraph 3: Commercial research

Option 0
no provision [i.e., using protected knowledge for commercial research without first obtaining the permission of the rights-holder is prohibited]

Option 1
The rights conferred in Article 14 are not infringed by the use of knowledge by commercial researchers for the purposes of developing:

Element 1
new plant varieties

Element 2
other innovations that may be derived from protected indigenous and local knowledge

Paragraph 4: Customary uses

Option 0
no provision [i.e., there are no exemptions, even for customary uses of knowledge by indigenous peoples and local communities]

Option 1

Notwithstanding any rights conferred under this Law, indigenous and local people shall be allowed to use indigenous and local knowledge as long as that use conforms to customary practices and uses.

Commentary

The options for rights conferred over protected indigenous and local knowledge cover a broad range of potential uses. This article includes options that would provide relief from those strictures. An alternative approach with same result would be for legislators to craft their 'rights conferred' more precisely in the first case. Then exemptions such as these would not be necessary. The manner in which we have presented rights, and then exemptions from those rights, is common in intellectual property legislation.

Paragraph 1 (Private use), Option 1 exempts natural persons from having to obtain the permission of the holders of protected knowledge as long as they are going to use that knowledge for their private purposes only. Not including this exemption would make this law more restrictive than many countries' patent laws, which allow for private-use exemptions. The European Patent Convention (EPC) Article 27 provides such an exemption. The US patent law (35 USC 100) does not.

Paragraph 2 (Educational research), Option 1 exempts the use of protected knowledge within an academic setting for research that is not commercially oriented. US copyright law includes 'fair use' exemptions for private and educational use. Most countries, however, do not have such exemptions. Fair dealing in Canada and the United Kingdom does not exempt all uses for private and educational purposes, and indeed, royalties are due to the collectives who manage such rights for most educational reproductions at a fixed tariff set by the government. Such tariffs may be lower for non-profit institutions, but they are not waived. There is no reason that indigenous peoples and local communities should not also benefit from the collective licensing regime in states where it is established and simply set different tariffs for various kinds of infringing uses.

Paragraph 3 (Commercial research), Option 1, Element 1 reflects the International Union for the Protection of New Varieties of Plants (UPOV) breeders' exemption, which allows plant breeders to use a protected plant variety for the purpose of developing another improved variety without asking permission. However, only the development is free, not any subsequent commercial exploitation of the variety developed, which may require the permission of the rights-holder.

Paragraph 3, Option 1, Element 2 is added because these provisions apply to a much wider range of indigenous and local knowledge than simply plant varieties. Again, only use of the knowledge for the development of the innovation would be free from the need to obtain permission; subsequent commercial exploitation of the innovation would require permission from the rights-holder.

Paragraph 4 (Customary uses), Option 1 has the advantage that it does not attempt to force customary uses into any one or combination of the other exemptions set out in Paragraphs 1–3. It may be, for example, that an indigenous group would not be comfortable with even a private use of their knowledge if it were not in conformance with customary practice. Reproduction of a sacred design on a T-shirt, even if is just for someone's private use, might not be acceptable. A customary-use exemption — one that allowed for the reproduction of the design in a more customary setting and medium, independent of whether or not it was private — would be more appropriate. One obvious disadvantage to this kind of exemption would be its uncertainty. Every community would have different customs. The advantage of the other exemptions is that they apply to all knowledge, in all situations.

Article 16: Duration of rights conferred

Paragraph 1: Same duration for all rights

The duration of rights conferred in Article 14 shall be:

Option 1
in perpetuity for all rights granted.

Option 2
for [one hundred years] [sixty years] [twenty years] [some of other period of time] for all rights granted.

Option 3
for as long as the rights-holder actively uses that knowledge. Once there has been a significant lapse in the customary use, the knowledge will revert to the public domain and all rights will expire.

Option 4
[perpetual] [certain number of years] for the original rights-holder. If the original rights-holder transfers those rights, the rights shall last only [shorter number of] years calculated from the date of transfer.

Paragraph 2: Different duration for different rights

Option 1

The duration of rights conferred in Article 14 shall last [for different periods of time depending upon the strength of the rights granted]:

Element 1

in perpetuity for integrity and attribution rights

Element 2

[in perpetuity] [some shorter period] years for non-exclusive rights

Element 3

[in perpetuity] [some period shorter than that granted for non-exclusive rights] years for exclusive rights

Paragraph 3: Different duration for different conditions of protection

Option 1

The rights shall endure:

Element 1

[perpetually] for knowledge that is kept confidential within the concerned group of knowledge-holders

Element 2

[perpetually] [less than in Element 1] years where novelty is a condition for protection

Element 3

[perpetually] [less than in Elements 1 and 2] [any number] years where originality is a condition of protection

Commentary

We do not purport to cover all of the possibilities in Paragraph 3, Elements 1–3 above. Instead, they are designed to illustrate the kinds of considerations that policy-makers must address.

The discussion about the appropriate duration of intellectual property protection for indigenous and local knowledge turns out to be a microcosm of all of the issues treated in this volume. The appropriate duration depends upon:

1 the purpose of creating a *sui generis* intellectual property law for indigenous and local knowledge in the first place (four such purposes are discussed in Topic 2, Section 1, A.1–A.4, above);

2 the scope of knowledge being protected (which depends upon both the definition of indigenous and local knowledge in Part Two and the conditions for protection in Part Three, above); and
3 the nature of the rights conferred (e.g., exclusive versus non-exclusive), as set out in Part Four, above.

For example, if one purpose of appropriating indigenous and local knowledge is that it will somehow preserve the cultural value of that knowledge to its holders, logic suggests that protection cannot ever lapse without some cultural cost to the community from which it came. Consequently, the rights conferred should be perpetual. On the other hand, a very different utilitarian principle underlies the establishment of time-limited rights in existing intellectual property laws. Theoretically at least, existing intellectual property laws embrace an acultural, utilitarian balance between (a) the necessity to create incentives to individual creativity, and (b) the benefit to the public of disclosing the nature of those creations (and eventually being able to copy them). Expiry of protection is a large (and the most public-minded) part of the balance struck by intellectual property law. On the face of it, the 'culturally integral' principle of infinite protection cannot be reconciled with the utilitarian principle of time-limited rights. Some sort of workable solution is crucial to the successful creation of a *sui generis* intellectual property law. Policy-makers will ultimately have to decide what duration of rights best expresses their overall objectives for the legislation.

At the same time, policy-makers should also consider the combined factors of (a) the breadth of the class of knowledge they would like to protect, and (b) the strength of that protection. If the conditions for protection were such that a great deal of not-very-specialized indigenous and local knowledge would qualify for protection, and the rights conferred in association with that knowledge were exclusive, then it might make sense to have a shorter duration. On the other hand, if the rights conferred with respect to that knowledge were relatively weak (e.g., non-exclusive and integrity rights), the duration of those rights could be longer.

Any limitation of the duration of rights implies the need to define the moment at which protection starts. This issue takes us back to Article 13, Paragraph 3, where the Crucible Group considers options about what knowledge-holders must be required to do to be entitled to rights. If registration is necessary, then the timing of the duration of the rights would start at registration. If registration is not necessary, protection should start as soon as the other conditions for protection are met. Another option is that protection should begin upon the first use of the protected knowledge by a third party.

Pursuant to most of the legislative options included in this collection, the knowledge that is protected cannot change once it has qualified for protection. While it is true that indigenous and local knowledge is dynamic and ever changing, the units or instances (for lack of a better term) of that knowledge that can be protected pursuant to these provisions are frozen at the time of protection. Any changes or incremental gains to that knowledge after it is made the subject of intellectual property protection would not be protected by the original grant of an intellectual property right. Such incremental gains would instead have to be made subject to subsequent applications (or whatever the provisions require) for protection.

If there are no administrative requirements for protection to vest in a community — as in the case of copyright law in many countries, which vests automatically upon the authorship of a work — such 're-application' would be largely fictional, as every original evolution in a work would automatically qualify for protection. The same could be said of knowledge that is protected on the basis that it has the quality of confidence. (For more on confidentiality, see Article 8, above.)

It has been argued by some that *sui generis* intellectual property protection should be explicitly constructed to cover knowledge that continues to change and evolve. In Viewpoint box 4, above, it is argued that inflexibility in protection contributes to the 'freezing' and eventual erosion of indigenous and local knowledge and cultures.

Viewpoint box 9: How long should protection rights last?

Forever!

Indigenous and local knowledge is inalienable; therefore, *sui generis* intellectual property controls over that knowledge must last forever. Anything less brings about absurd results: 20-year protection for knowledge that has been accumulating for thousands of years.

The duration should be subject to utilitarian considerations.

Indigenous and local knowledge is inalienable. But this principle has no implications for the period of intellectual property protection except by way of a poor analogy. After the period of protected exclusive use is expired, no one is going to take knowledge away from indigenous and local communities. The principle of inalienability would still be respected after the protection expires. Consequently, the duration of intellectual property protections for indigenous and local knowledge can also be subject to utilitarian considerations. For example, national laws to protect indigenous and local knowledge should also have the goal of maximizing knowledge flow. Therefore, the duration of protection should be inversely proportional to the scope and strength of protection.

Furthermore, non-indigenous and non-local communities do not exist in cultural and spiritual vacuums. They too have knowledge integral to their identity. Why, then, should indigenous and local communities get unlimited protection when others do not?

Rights should endure for as long as the knowledge is used by the community (a middle ground).

Community rights of control over their knowledge should last as long as the community use of that knowledge is active.

Paragraph 1 (Same duration for all rights) includes three options, each for a different period. What unites these options is that all of the potential rights that might be included in the legislation would have the same duration, regardless of the nature of the knowledge protected, the strength of the rights granted, and so on.

Paragraph 2 (Different duration for different rights) includes options wherein the duration of rights conferred would vary depending upon the strength of the rights conferred.

Paragraph 3 (Different duration for different conditions of protection) includes options wherein the duration of rights conferred would vary depending upon the conditions of protection that are eventually included in the law.

Article 17: Compulsory licences

Option 0

no provision [i.e., there are no conditions under which compulsory licences can be granted]

Option 1

The competent authority may order owners of indigenous and local knowledge to grant licences to third parties to use indigenous and local knowledge when:

Element 1

the rights-holder abuses its rights [e.g., sets prices artificially high, enters into anticompetitive intellectual property rights pooling with other intellectual property rights-holders, etc.]

Element 2

the rights-holder fails to make available an adequate supply of the knowledge, through its own supply, or through licensing, in situations where the competent authority decides that there is an overriding public interest. These situations may include:

a) national emergencies;
b) environmental issues;
c) public health problems; or
d) other issues.

Commentary

Compulsory licences are issued as remedies in those situations where a competent authority determines (1) that the intellectual property owner is abusing his or her exclusive rights, or (2) that, although the intellectual property owner is not abusing his or her rights, the public interest in availability of a protected technology outweighs the intellectual property owner's interest in his or her exclusive rights.

Compulsory licences are not relevant if the rights conferred in association with the protection of knowledge are non-exclusive.

Some Crucible members feel that we should not even be including a section on compulsory licences, because to implement these options would be to create a weapon for the strong to use against the weak. Others feel, however, that once one subjects a new class of knowledge to intellectual property rights — whether the party benefiting from that protection is weak or strong — it is absolutely necessary to provide the government with a 'way out' to make sure that the public interest is at least minimally protected against potentially hurtful practices by intellectual property owners.

Article 18: Anticompetitive practices

Commentary

Another potential restriction on the freedom of intellectual property owners to exercise their exclusive rights comes in the form of anti-trust or competition laws. Many countries' competition laws have provisions to preclude intellectual property owners from 'pooling' their patents, thereby creating monopolistic cartels or arranging exclusive supply agreements among themselves, etc. The Crucible Group will not analyze options for these sorts of laws in this volume.

Part Five: Procedural/administrative matters

Article 19: Licensing agreements

Paragraph 1: Approval

Option 0

no provision [i.e., only the rights-holder needs to approve]

Option 1

The [competent authority] shall be notified of a proposed licensing agreement regarding the use of knowledge protected under these provisions.

Licence agreements for the use of indigenous and local knowledge will be null and void unless, in addition to the knowledge-holders, the [competent authority] also provides its approval.

The [competent authority] shall grant approval of the agreement within [number] months provided the agreement complies with the minimum conditions set out in Paragraph 2 of this article.

Option 2

The [competent authority] shall, if requested to do so by an indigenous or local group, provide that group with information, assistance and expert advisers for the purpose of carrying on a licensing negotiation under this Act.

Paragraph 2: Conditions of approval

All licence agreements between rights-holders and licensees [must as a minimum] [may] specify:

Option 0

no provision [i.e., no specified conditions]

Option 1

Element 1

complete details about the corporate/institutional history of the licensee

Element 2

complete details of the research the licensee is engaged in

Element 3

the intended use of the indigenous and local knowledge the licensee seeks

Element 4

the nature of the legal rights the licensee may seek over innovations based on the protected knowledge in this country and abroad

Element 5

minimum terms regarding benefit-sharing [as laid down in government order]

Element 6

that specimens of all biological material shall be deposited with an institute for the deposit of such material within this country

Element 7

that all indigenous and local knowledge collected shall be summarized in written form and deposited with an institute based in this country

Element 8
that any technologies derived from biological material and/or local knowledge collected from a particular community shall be made available for use by that community [on an unrestricted basis] [at a prorated cost taking into consideration the cost of production] [at a prorated cost taking into consideration the market value]

Element 9
that some proportion of the research undertaken by the applicant shall be conducted in partnership with members of the relevant community or with an organization associated with the community

Element 10
that the licensee shall include information regarding the source of, and extent of reliance upon, indigenous and local knowledge collected pursuant to the agreement in any applications the applicant makes for intellectual property rights for technologies derived from such biological materials and local knowledge

Element 11
that the licensee will undertake to train local people

Element 12
that the licensee will make any technologies that make use of the knowledge provided by the rights-holder available to the rights-holder under preferential terms

Paragraph 3: Publication of application

Option 0
no provision [i.e., no need to publish the application]

Option 1
The licensee shall publish a written account of the fact that he or she is seeking permission to use knowledge protected under this Law. The publication shall include details about the particular area, communities that might be involved, the end towards which the research is being conducted, and the identity of all of partners who are jointly involved in the use of the protected knowledge.

Paragraph 4: Consultations

Option 0
no provision

Option 1

The [competent authority] shall hold [public] consultations with interested parties regarding the proposed use of indigenous and local knowledge.

Commentary

This article addresses the approval of licence agreements regarding knowledge protected under this law. Readers will notice that there is a great deal in common between negotiating a licence for use of knowledge protected pursuant to these provisions and negotiating agreements for access to biological resources pursuant to the access laws. The list of substantive requirements in Paragraph 2 are very much like those that applicants must negotiate to get access pursuant to the Philippines' EO 247, Andean Pact Decision No. 391, the Indian draft Biodiversity Conservation Act, and those set out in Topic 1, Section 2, Articles 7, 8 and 9 of this volume.

Paragraph 1 (Approval) addresses the question: who should be allowed to consider requests to use protected indigenous and local knowledge? The rights-holders, certainly. It has been suggested, however, that for the purpose of protecting the interests of indigenous and local communities, a government authority should also be involved in the consent-giving process.

The options under this paragraph are very similar to those listed in Topic 1, Section 2, Article 6. Many laws to regulate access (e.g., Philippines' EO 247, Lao draft access legislation, Indian draft Biodiversity Conservation Act, Andean Pact Decision No. 391, Organization of African Unity (OAU) Draft Model Provisions) require the consent of the national government, in addition to local communities, when foreigners apply for access to indigenous and local knowledge and biological resources on indigenous and local lands.

Paragraph 2 (Conditions of approval) addresses the possibility of having mandatory substantive terms to be included in all licence agreements concerning protected indigenous and local knowledge. Another alternative, of course, would be not to include any such terms, and to allow the parties total discretion to arrive at whatever deal they choose. A middle position would be to require the parties to go over a form wherein they indicate that they have considered each listed option. They would not be required to include each term in their agreement, but they would be required to indicate that they had considered the term. This would at least protect indigenous and local negotiators from not understanding the kinds of benefits they might be able to negotiate for. Most other intellectual property laws do not include such detailed provisions to regulate licensing processes and terms. However, the competition law of many countries prohibits certain clauses to be included in technology-transfer agreements that could lead to unfair competition and restraint of trade.

In the absence of mandatory conditions, unequal bargaining power may leave indigenous and local communities settling for less than they might otherwise have gained out of a deal if the conditions had been mandatory in the first place. Clearly, there is a danger that if the law requires too much from the party seeking to use knowledge, that party simply will not use the indigenous and local knowledge.

Policy-makers might want to maintain flexibility regarding which of these terms must be included in each contract. Flexibility is needed to accommodate the fact that the significance of the contribution of the local knowledge to the development of derivative technology will be different in each case. Policy-makers might wish to include more beneficial terms for the community in cases of a larger contribution. Flexibility such as this would require the creation of an objective test for remoteness.

Viewpoint box 10: Should the government be involved in the licensing process?

Of course.

Given the imbalance of power between indigenous and local communities and the multinational parties that use indigenous and local knowledge, the government's involvement is essential to make sure indigenous and local communities' interests are not trampled. The role of the government in approving or disproving others' use of indigenous and local knowledge would be limited to making sure indigenous and local communities are not ripped off. In many countries, the national government already plays this role with respect to proposed sales of aboriginal community lands. Some would argue that the CBD only requires national governments' PIC, and not the community's consent. So communities are lucky to be included at all! A logical extension and ancillary benefit to incorporating governments into the consent-giving process would be an obligation on governments to chase down violators who use protected indigenous and local knowledge without first seeking permission. Enter the 'national indigenous and local knowledge protection ombudsman'?

No way.

Why should indigenous and local communities have to get their national government's permission to do what they want with their protected knowledge when patent, copyright and Plant Breeders' rights-holders do not have to do the same thing? Why should this newly evolving area of law embrace the same old paternalism with respect to indigenous and local community interests? It is true that existing national access laws integrate the government's PIC in addition to that of the local communities. But that is a problem with access laws. They are structured to the advantage of national governments. It would be better for communities if governments were not required to interfere with their deals.

Maybe, but just a little bit.

Government's participation would be acceptable if its role were legally restricted to last-minute, non-intrusive reviews of all-but-finished indigenous and local knowledge licensing agreements to make sure communities were not getting significantly less than the market value of their knowledge. If this option were coupled with the creation of an indigenous and local knowledge protection ombudsman, it might be more appealing.

Another alternative would be to have government involved only upon the request of a deal-making group requiring assistance.

Paragraph 3 (Publication of application), Option 1 requires the licensee to publish a written account of the fact that he or she is seeking permission to use knowledge protected under this law. This requirement may help to identify potential rights-holders, which is particularly important if there is no registration system for indigenous and local knowledge.

Paragraph 4 (Consultations), Option 1 requires consultations to be held with interested parties. This may be helpful since the use of indigenous or local knowledge may also be of interest to those who are not the rights-holders. Of course, this condition may also act as a disincentive for bioprospectors. Often they do not want to make a public disclosure of what they are looking for and why. To do so would be giving their competitors valuable information. In the end, forcing them to make a public disclosure as a precondition to obtaining permission may completely undermine their interest in access to the resources in question.

Article 20: Ombudsman's office

Option 0

no provision [i.e., there is no ombudsman's office]

Option 1

Element 1

The holders of protected indigenous and local knowledge may notify the ombudsman of situations wherein they believe their knowledge is being used by third parties without their consent.

Element 2

The ombudsman may conduct an independent investigation into whether or not protected indigenous and local knowledge is a component of the party's activities. Where the ombudsman informs the party that their research activities involve protected indigenous and local knowledge, the ombudsman may further recommend to that party that they should seek permission, in the form of a licence agreement, with the rights-holders [and the competent authority] prior to using that knowledge. Alternatively, if the rights violated are non-exclusive rights, the ombudsman may recommend to the party that they should be making proper attribution, or remuneration, to the knowledge-holders.

Element 3

If the researching party disregards the ombudsman's recommendation, the ombudsman may refer the case to the proper dispute-resolution or decision-making body. In the case of such a reference, the ombudsman will notify both the rights-holders and the parties using the knowledge that he or she has referred the case to the competent authority.

Element 4

The ombudsman may undertake to make representations to the dispute resolution or decision-making body on behalf of the indigenous and local knowledge-holders.

Commentary

This option does not create an ombudsman with jurisdiction to decide cases of alleged illicit use of indigenous and local knowledge. It does, however, involve the dedication of resources to assist indigenous and local parties to enforce their rights pursuant to *sui generis* intellectual property laws. The ombudsman could help indigenous and local knowledge-holders discover cases where their rights have been infringed. The ombudsman could then provide badly needed advice to rights-holders regarding the most appropriate forum in which to seek remedy. Finally, the ombudsman could assist the aggrieved parties to prepare and present their case before the relevant dispute resolution or decision-making body.

Exactly what judicial or administrative forum would be most appropriate in each case would depend upon a variety of factors. If the aggrieved parties' first interest was to block the violators from getting intellectual property rights themselves, then the first forum they could go to would be the intellectual property offices themselves. For example, the national plant variety protection law or patent law could require applicants to show proof of having obtained the PIC of communities whose knowledge they have used to develop their inventions or plant varieties. In the material in this volume regarding national plant variety and patent laws, we include options for exactly these kinds of provisions. Where the community can establish that their

Recommendations

1 Given the technical difficulties associated with enforcing *sui generis* intellectual property protections for indigenous and local knowledge, countries attempting to implement such provisions should create an ombudsman's office, either to take the lead, or at least to provide assistance to help indigenous and local communities to enforce their rights.

2 Given that much of the knowledge concerned in these provisions moves across international borders, countries that are serious about creating *sui generis* intellectual property protection for indigenous and local knowledge should lobby for the creation of international agreements or cooperation treaties to harmonize *sui generis* intellectual property standards in different countries.

knowledge was used without their PIC, the patent or plant variety protection office could refuse to grant intellectual property rights in favour of the applicant.

Second, in cases where knowledge-holders wanted some other form of remedy, they could go to civil courts to petition for compensation, prohibition orders, and so on.

Third, if the *sui generis* law creates criminal or quasi-criminal offences out of illicit use of indigenous and local knowledge, the knowledge-holder could complain to state authorities, who could then decide whether to prosecute the offender.

Alternatively, *sui generis* legislation could include provisions creating a specialized tribunal to hear cases of violations. It would be premature at this point to enter into a detailed analysis of this tribunal.

Article 21: Competent authority

Option 1

The [Patent Office] [Plant Variety Protection Office] [newly created Office for the Protection of Indigenous and Local Knowledge] shall function as a competent authority under this Law.

Commentary

New *sui generis* intellectual property provisions could be administered by either (a) an existing body or (b) a newly created body, within the country implementing the provisions. The plant variety protection or patent office could add *sui generis* intellectual property laws for indigenous and local knowledge to its roster of responsibilities.

It is necessary to be cautious when considering the creation of new government bodies to administer new laws, especially in countries where there is already a great deal of corruption in the civil service. The new offices of an ombudsman and competent authority could become the locus of a network of corruption involving officials and companies wishing to do business with indigenous and local people. The stakes for the control of some indigenous and local knowledge would be very high. There would be massive incentives for poorly paid officials in developing countries to take advantage of their official positions. The creation of rights over biological resources is characterized by legal complexity, the transplant of alien property institutions and the creation of power over vital resources — all of which are breeding grounds for corruption and fraud.

Part Six: Remedies

Article 22: Offences

Commentary

As discussed in the commentary to Article 20, above, numerous potential causes of action and remedies flow from the options presented in the previous parts. Most importantly, the options for laws presented in Parts One to Five set the foundations for civil actions, criminal prosecutions, and patent and plant variety protection interventions in response to alleged infringements on the rights of indigenous and local knowledge rights-holders. The range of potential remedies (depending upon the body making, arbitrating or deciding the case) is also very wide, and could include prohibitions, compensation, fines, restitution, probation, jail terms, and/or refusal to grant, or revocation of, patents or Plant Breeders' Rights.

Article 23: Defences

Commentary

Just as the options for law in this collection create a legal foundation for causes of action and potential remedies, so too do they create the legal groundwork for possible defences. For example, if the *sui generis* legislation included the requirement that indigenous and local knowledge had to be registered before it could be protected, a possible defence against an allegation of infringement would be that the alleged infringer had conducted a reasonable search of the registries and found no relevant registered knowledge.

Part Seven: Relationship to other laws

Article 24: Plant variety protection and patent laws

Option 0
no provision [i.e., double protection is possible]

Option 1
Subject matter protected under this Act may not fall under the protection of any other intellectual property regime.

Commentary

Part Seven considers the relationship of these *sui generis* intellectual property laws for indigenous and local knowledge to other laws, in cases of overlapping subject matter. There are two main options: to make no provision, or to provide that knowledge protected under these provisions has priority. A third option exists logically, but probably not practically: that is, to give priority to other forms of protection. However, this would be inconsistent with most reasonable objectives for the present law, so we do not consider it further.

Option 0: The normal rule for construing legislation is that independent laws do not contradict each other. Laws are construed to be mutually consistent — except where this is clearly not possible, or where one law specifically refers to and supersedes another. Likewise with rights. In intellectual property law systems, subject matter (or aspects of it) may be protected by one or more patents, by a registered design, by copyright (for example, in the design of its label) and by trade marks affixed to it. None of the rights necessarily interferes with any of the others, so all can be given effect. In the case of protected indigenous and local knowledge, the pre-existence of the knowledge as such would generally prevent subsequent valid patents on the same knowledge. It would not prevent patents on developments of such knowledge. Depending on the form of the rights conferred in the indigenous or local knowledge, the grant of patents on innovations building on that knowledge would not nullify the rights of the indigenous or local knowledge rights-holders; the latter would be able to assert their rights against the patent-owners or their licensees, to control or prevent exploitation or to recover royalties.

Option 1 subordinates the rights of other intellectual property owners to the rights of the 'owners' of traditional knowledge. This subordination might be interpreted narrowly or broadly. Narrowly interpreted, this option prevents only the grant of intellectual property rights over indigenous and local knowledge in the form in which it is already protected under the present provisions. As noted, such later rights are likely to be invalid for other reasons (for example, lack of novelty). If not, a provision reinforcing this effect (to invalidate or render ineffective intellectual property rights granted over such knowledge) would be uncontroversial. However, a broad interpretation of this section would also prevent any patenting of genuine further innovations building on protected indigenous and local knowledge. Many instances of so-called 'biopiracy' are exactly this. While this option would be effective to dispose of such instances of 'biopiracy', it would equally interfere with protection, under other intellectual property systems, of any innovations using traditional knowledge, even with the consent of the owners of the knowledge. Broadly interpreted in this way, this option would fundamentally conflict with TRIPs, as an unjustified restriction on intellectual property protection.

Topic 3

Options for national laws regarding biological innovations

Contents

Section 1

Survey of domestic policies to promote biological innovations

As governments struggle to encourage innovation in science and technology, it is helpful to look to history to try to understand the successful tools and tactics of the past. Science and economic historians tend to emphasize the Industrial Revolution in Europe. Three broad influences are often cited:

- **Environment** — although the geo-environmental place of Europe is often identified as significant, this was only part of the force that created a culture of 'tinkers' or of practical enquiry. The Industrial Revolution arose in a welcoming sociopolitical environment. In today's world, this might translate into government policies that are empathetic to experimentation.
- **Education** — while the innovative environment led to an emphasis on formal education, it was the move to technical training, especially in Continental Europe, that provided the basis for sustained technological progress. Today, this translates into support for higher education.
- **Entrepreneurship** — it was realized that technological advances could directly improve local (national or corporate) productivity. New ideas that reduced production costs or increased markets were quickly internalized.

It is often argued that governments had a rather small role in the Industrial Revolution and that scientific innovations came from individuals or private enterprise. Yet the innovators themselves benefited from public and private education. Government grants, awards or subsidies also supported innovative industries and individuals. Whether it was for the improvement of ships and navigation in the British Admiralty or for the development of a steel industry in France, governments were not averse to offering awards to successful inventors for developing a technological solution to a problem.

During the Industrial Revolution, innovators were also supported by defined national goals such as those occasioned by the Napoleonic Wars, or the military urgency of improving steel or developing chemicals. In such efforts, governments employed numerous devices from subsidies to monopolies in order to encourage rapid technological advances. In this, the times have not changed. In the 1940s, the Manhattan Project (the project to build the first nuclear bombs) cost American taxpayers US$2 billion, but the scientific information it generated served to fuel the entire commercial nuclear power industry. Without direct government intervention, it is arguable that this industry would not exist today.

National purpose — especially the Cold War — also provided the necessary funding for scientific breakthroughs in aviation, telecommunications and data management. Improvements in air transport in the 1950s, and the space programme of the 1960s, facilitated by government expenditure on defence and aerospace research, led to the creation of whole new industries. Public funds for military satellites have evolved into a new telecommunications industry with commercial satellites. Many of the breakthroughs in microelectronics can be traced back to the space programme and defence-funded research. Advances in materials science have the same origin. The Internet began in 1966 as a defence project.

Advances in agricultural productivity in the last century have little connection to military spending. Nevertheless, many of the major breakthroughs in yield and farming systems have come from the public purse. Agricultural universities — most of them created in the early 20th century — dominated most crop improvement. The 'Green Revolution' was an international public-sector research function. At least some of the advances in plant breeding can be credited to developments in computer data analysis. The initial breakthroughs in biotechnology — whether the discovery of DNA in the 1950s or gene manipulation in the 1970s — took place in universities and government-funded laboratories. Many of the major advances in medicine were publicly funded. The Human Genome Project began as a US$3 billion international public research endeavour. Developments in computer sciences and biotechnology are now slashing costs and bringing the work within reach of private-sector funders. Most of the research investment in novel areas such as nanotechnology continues to come from taxpayers.

The spin-off effect of 'mega-science' has created many new industries and often unanticipated economic benefits. In formulating agricultural research policies, governments should bear the past and present history of innovation in mind and be prepared to exercise the full range of tools available to ensure that the public good is served. Chief among these tools continues to be education and the stimulation of an inquiring environment.

While the foregoing discussion speaks of the need for sound policies to stimulate education and an innovative environment, it does not address the role of entrepreneurship, either by individuals and communities, or in the private and public sectors. Entrepreneurial innovators work best (or only) if they can gain directly from their investment of time and/or money in research. It has often been argued that 70% or more of innovations in the private sector come about because the innovation directly saves the enterprise money or increases its market share. If this is so, government policies should be designed to inform and assure innovators that the regulatory environment welcomes innovation.

At this point, most private-sector agricultural researchers would insist that adequate intellectual property protection is essential to commercial innovation. There is a wealth of anecdotal evidence to suggest that this is so. There is not, however, much empirical data to support this contention. Partly this is because of the difficulty in isolating the incentive created by 'patents' from other incentives such as those discussed above. After more than two decades of intense international debate, the empirical basis for intellectual property's incentive value remains unclear. In considering legislative mechanisms to support innovation, policy-makers should note that intellectual property law could be seen as one possible (and controversial) instrument among more than two dozen fields of state intervention. Some Crucible members do regard it as an important one, however.

Support for higher education and public breeding rank among the most practical steps governments can take to encourage domestic innovation in agricultural research. This is especially important for developing countries, which may lack markets rich enough to attract commercial investment in plant breeding. Stability in research funding is also particularly important. Some university researchers emphasize the importance of broad training curricula: they believe the trend to specialization and R & D contracts with private-sector enterprise can undermine full-spectrum training. Contractual obligations that could restrict academic freedom — particularly the ability of a teacher to educate students — should also be avoided.

Award systems in the public sector can vary from simple recognition, to increased equipment and human resource support, to personal financial benefit. In many countries, governments make it possible for public researchers to benefit directly from intellectual property grants.

Farming communities benefit from the whole spectrum of government programmes and policies for innovation. They may gain most from a sympathetic policy environment that favours secure land titles, encourages rural credit under reasonable conditions and maintains accessible extension services geared to local needs. Small-scale, resource-poor farmers also welcome policies that encourage entrepreneurship. Governments can promote agrobiodiversity by working to ensure that local and national markets do not erect artificial packaging, labelling or health barriers biased to large commercial growers and importers. Governments can promote diversity, in fact, by supporting pro-diversity labelling and through public education campaigns that attract local consumers to local produce. Seed certification schemes can be reviewed to ensure that they are not unduly restraining local diversity. Again, pro-diversity seed labelling could be considered. Since genetic diversity in the field can pose special problems for seed cleaning and milling, governments may wish to work with local entrepreneurs and/or farmers' cooperatives to overcome some of the purely mechanical barriers that prevent diversity from reaching markets.

Governments can also encourage cooperative research partnerships between farmers and the public and private sectors. Incentive systems could draw public breeders to work more closely with local communities. Similarly, communities can support useful public breeding initiatives through farm 'check-off' programmes (where a small portion of market profits are volunteered back to the public breeding work) or through land and labour contributions to research both sides consider valuable.

Governments can also facilitate local plant breeding by recognizing the complex and unique characteristics of community-led research. Unlike most of their institutional counterparts, small-scale farmers act simultaneously as breeders, growers and the primary consumers of their own research and production. In keeping with most farmers around the world working with most non-hybrid crops, small-scale farmers save seed from each harvest for planting the following season.

Farmers in developing countries routinely select-out mutant, high-yielding, or otherwise high-quality seed for testing and experimentation. While institutional breeders, whether public or private, are not in a position to breed for the whole range of high-stress conditions faced by subsistence farming communities, such farmers (mostly women) regard harvested seed as their initial source of variation from which they can undertake their own plant breeding. Interesting material is tested in tiny plots adjacent to the home. Often, promising germplasm is exchanged with neighbours and tested against different slopes and soils in the community. The most promising material is multiplied and incorporated into the seed stock for the next planting.

Community plant breeding can produce radical changes in the field, but it is more likely to be a gradual process that allows the diversity of farmers' varieties in the field to evolve over time. Nevertheless, the germplasm in the field does change somewhat every year, and yield and quality improvements are recorded in various ways by the farmer-breeders themselves. Some advocates insist that the distinctions that arise from year to year often match or exceed the distinctions found within a 'family' of successive varieties released by an institutional breeder.

An important element in community plant breeding is the exchange of 'exotic' germplasm among farming communities. Farmers are always eager to exchange seed. Every country and culture can point to a long history of customary seed exchange — sometimes linked to religious practices and other times associated with markets or celebrations. The exchange process is a major opportunity for introducing significantly new seeds. It is this process that sped maize through many of the growing regions of Africa in a few generations and allowed sweet potatoes to spread through East Asia and the Pacific in less than a hundred years.

The kind of germplasm exchange that takes place in local or regional markets can be paralleled to the transactions that take place between genebanks, except that those exchanging seed in the community are exceedingly poor. As genebanks sometimes seek to recoup their costs by charging for seed multiplication and shipment, farmer-breeders recoup their costs, to the extent possible, through direct exchange, barter or cash. Regardless of the type of transaction, the objective is exchange for the purposes of crop improvement.

Governments wishing to support community plant breeding should adopt policies and practices that encourage this system and make sure that other regulations or legislation do not work to curtail this kind of innovation. There may be general agreement that three factors, environment, education and entrepreneurship, provide a sound basis for sustainable innovation. However, the exact policies to be pursued within these areas are much in dispute. History suggests that governments should have a sense of direction — some clear notion of the kinds of innovations that could prove most socially useful. History also suggests that governments do well to invest in research and development. Many of the key incentives may require little or no investment but are more in the form of stimulating the research environment and removing unintentional barriers.

To conclude, there are many ways for governments to stimulate agricultural innovation. The following list summarizes possible policy mechanisms, some of which have already been discussed in the above paragraphs. Although the discussion in this section makes specific references to innovation in plant breeding and agriculture in general, the policy initiatives mentioned are equally relevant for stimulating innovation in other areas, such as pharmaceuticals.

Policies generating knowledge:

- Support for agricultural training
  - secure university funding
  - secure broader training in plant breeding disciplines
  - secure academic freedom
- Encouragement of research and development
  - award systems (one-time cash or certificate awards)
  - public plant breeding
  - intellectual property legislation
- Encouragement of research partnerships
  - farmer cooperatives for plant breeding
  - farmer 'check-offs' (voluntary contributions) for public breeding
  - other cooperative research initiatives (public and private)
- Support for indigenous and local knowledge (see indigenous and local knowledge section)

Policies generating opportunity:

- Encouragement of infrastructure
  - land-tenure security
  - credit access
  - transportation infrastructure
  - taxation and subsidy incentives
  - public agricultural extension support
  - plant breeding for minor crops and marginal lands
- Promotion of market diversity
  - seed fairs (promoting local diversity)
  - labelling (celebrating local diversity)
  - consumer protection regulations (monitoring provisions that could constrain diversity)
  - certification regulations (making it easy to grow and market diverse seeds)
- Promotion of competition
  - anti-trust legislation
  - 'anti-gatekeeper' legislation (preventing monopolies on core technologies)
- Support for germplasm access
  - conservation (ensuring that diversity is not lost)
  - exchange (ensuring access to genebanks and farmer-to-farmer exchanges)
  - research exemptions (encouraging community and institutional plant breeding)

Section 2

Options for *sui generis* intellectual property laws for plant varieties

Contents

Section 2

Options for *sui generis* intellectual property laws for plant varieties

Part One: Purpose and scope

Article 1: Purpose

The purpose of these provisions is to:

Element 1
make plant varieties subject to private property rights

Element 2
ensure that plant breeders are able to recover costs, make sufficient profits from useful innovations and continue to innovate

Element 3
provide legal recognition for varieties that are not protectable under existing patent and/or Plant Breeders' Rights laws, thereby recognizing the value of farmers' plant variety innovations; and to provide a means of sharing the benefits derived from the use of farmers' or traditional varieties as breeding material and/or for commercial purposes

Element 4
encourage innovative plant breeding

Element 5
encourage the development of technologies that support sustainable farming practices

Element 6
conserve and increase plant species diversity and plant variety diversity

Element 7
promote global food security

Commentary
Many acts have no preamble or statement of purpose. The elements listed here are the most likely purposes or objectives of the legislation. Of course, not all the various possibilities are included here, but readers will find the most common justifications for creating intellectual property rights for plant varieties. An explicit statement of purpose is, however, no guarantee that objectives will be reached. This can be ensured only through the appropriate choice from among the tools provided in the following sections.

The purposes set out in Elements 1 through 7 are arranged from the narrowest and most mechanistic (i.e., making plant varieties subject to private property rights) to the widest and most ambitious in scope (i.e., promoting global food security). Crucible Group members' comments on this list of elements show two interesting trends. First, there is increasing agreement among the members about the utility of the purpose identified. For example, not everyone agrees that there is any utility in subjecting plant varieties to private property controls (Element 1), but everyone agrees with the purpose of promoting global food security (Element 7). Second, there is a decreasing level of agreement within the group about the ability of intellectual property rights to actually advance the identified purposes. Everyone agrees, for example, that intellectual property laws create 'owners' with property rights in plant varieties (Element 1) and that those rights contribute to plant breeders' ability to recover their costs and make profits (Element 2). They are much more divided over whether or not those rights promote sustainable farming practices (Element 5), plant species diversity (Element 6) or global food security (Element 7).

Article 2: Scope

Option 1
This Law shall be applied to plant genera and species as specified in the 'List of genera and species'. The competent minister is authorized to establish and amend the list.

Option 2
This Law shall be applied on the date of its coming into force to at least five plant genera or species and shall,

a) by the expiration of a period of three years from the said date, include at least ten genera or species in all;

Viewpoint box 11: Are intellectual property laws for plant varieties worthwhile?

Intellectual property law fails to advance issues of wider public interest.

Intellectual property laws for plant varieties create a form of private property where none existed before. In so doing, they provide an infrastructure for plant breeders to make a profit.

But no form of intellectual property legislation involving exclusive monopoly can address the needs or protect the interests of farming or other rural communities.

There is no evidence in the history of Plant Breeders' Rights, for example, to indicate that this form of intellectual property protection has ever encouraged credible agricultural research, stimulated R & D investment, or led to an increase in or diversification of the number of species or breeders involved in plant breeding. Indeed, the evidence suggests the opposite. To reinforce existing intellectual property laws regarding plant varieties, and to pursue new forms of so-called 'pro-farmer' legislation based upon this failed Plant Breeders' Rights experience, is to further jeopardize the food security and well-being of 1.4 billion people.

Intellectual property law does advance loftier public interests through the mundane mechanism of privatization.

On the contrary, the last 70 years of experience with plant variety protection have proven just how widespread the benefits of privatization can be.[22] Plant breeders enjoying the legal infrastructure of plant variety protection have created thousands of new varieties with dramatically higher yields. Those yields have paid off in dividends for farmers who grow the new varieties and in lower food costs for the public in general.

There is also evidence that institutional plant breeding has actually increased the level of plant genetic resources for food and agriculture (PGRFA).

Given the proven track record of plant variety protection laws, it is definitely worthwhile to explore whether or not similar benefits can be realized through the extension of modified laws to farmers' varieties and local farming communities. Providing local farmers with controls over their varieties, and a right to a share in the benefits of others' use of them, can only encourage those farmers to continue their socially useful innovations. Such laws would also provide a long overdue boost to the image of farmers as breeders and researchers.

b) by the expiration of a period of six years from the said date, include at least 18 genera or species in all; and
c) by the expiration of a period of eight years from the said date, include at least 24 genera or species in all.

Option 3

This Law shall be applied on the date of its coming into force to all species and genera.

Commentary

Although the law is necessarily concerned with plant varieties of different species and genera, countries have a wide range of options in selecting plant species and genera to protect. The options range from the possibility of protecting a limited number of species or genera to protecting all species and genera. The options also comprehend the possibility of vesting protection in these species or genera immediately upon the enactment of the legislation, or over time according to a graduated schedule. We do not claim to have discovered an optimum rate of additions in Option 2. It could easily be that some other rate would be appropriate, given differences among countries. The schedule included here comes from the International Union for the Protection of New Varieties of Plants (UPOV) Agreement 1978.

Countries choosing Option 1 would have to specify plant species and genera in a list. However, they should note that Article 27.3(b) of the Trade-Related Aspects of Intellectual Property Rights (TRIPs) Agreement requires the "protection of plant varieties by patents or by an effective *sui generis* system or by any combination thereof". As TRIPs does not qualify species or genera, it would appear that World Trade Organization (WTO) members must offer protection for plant varieties of all species and genera, or else this option may not be considered TRIPs-compliant. WTO member states may violate their TRIPs obligation if they decide to implement the minimum requirement of UPOV 1978 by granting protection for varieties of a limited number of species or genera only.

The elements enumerated here may be characterized as 'free choice' (Option 1), 'UPOV 1978' (Option 2), and 'no choice' (Option 3). Under Option 1, countries could freely define the specific number of species and genera falling under this law. Option 2 specifies the minimum number of species and genera for which protection must be made available and the relevant time-frame. Pursuant to Option 3, countries would not list specific plant species and genera; the law would cover them all.

Article 3: Exclusions from scope

This Law shall not be applied to plant varieties:

Element 1
for which patent protection is available

Element 2
which have merely been discovered

Element 3
which have merely been discovered in the wild

Element 4
the prevention of the commercial exploitation of which is necessary to protect *ordre public* or morality, including the protection of human, animal, or plant life or health, or to avoid serious prejudice to the environment

Element 5

within a genus or species with a particular manner of reproduction or multiplication, or a certain end use. The competent authority may limit the application of this Law to such plant varieties

Commentary

Obviously, the range of material that will be excluded from protection by this law depends upon the scope of the law. As in Article 1, there is a wide gulf separating those who think conventional UPOV-style legislation is adequate and those who recommend alternatives with more restrictive or wider forms of protection. Those questioning the need for any form of protection argue that wider categories of exclusion are necessary, for example, for plant varieties considered vital to national food security or developed using diversity-threatening technologies. Others insist that such exclusions should be on a case-by-case basis and that each case should be both verifiable and compensated for.

As an alternative to exclusion from the scope of the law, one might consider including special provisions concerning these varieties. For example, one might grant the discoverer of a variety different rights from those granted to the breeder of a variety.

Element 1: This exclusion is based on the idea that a variety should not be protected by both Plant Breeders' Rights and patents at the same time. It is based on and reflects the so-called ban on double protection as foreseen in UPOV 1978, Article 2.1. It aims at avoiding situations in which certain uses of a plant variety could be freely permitted under this law, but prevented at the same time by other conflicting intellectual property rights titles.

Elements 2–3: The key word in these two elements is 'discovery'. The exclusion in Element 2 is based on the idea that the mere discovery of a plant should not result in any exclusive rights of the discoverer with respect to this plant. Element 3, however, emphasizes discoveries in the wild. To understand these elements, it helps to recognize the essential difference between patent law and the UPOV Conventions. The UPOV Convention aims to secure the development of agriculture. The discovery, therefore, of mutations or variants in a population of cultivated plants is a source of varieties of great economic importance for agriculture. The UPOV Convention of 1978 was a special *sui generis* system that encouraged all forms of plant improvement, including discoveries (in contrast to the patent law, which did not protect discoveries). While it contains no express reference to the protection of discoveries, this protection can be inferred from the fact that the opening words of Article 6.1(a) accept the possibility that the variety may result from a natural source of initial variation, for example, a mutation.

When the Convention was revised in 1991, it was thought to be useful to include a definition of 'breeder' to emphasize the Convention's protection for varieties that had been 'discovered'. However, the apparent protection of bare discoveries could be controversial, particularly in relation to the definition of ownership rights in genetic resources. Nevertheless, it was recognized that making discoveries available is an important source of plant improvement that must be encouraged by the UPOV Convention and that, in practice, a discovery must be evaluated and propagated before it can be exploited. The resulting discussions led to the definition of 'breeder' as the person who "bred, or discovered and developed" a variety. According to one interpretation in the 1991 Convention, 'discovery' describes the activity of "selection within natural variation", while 'development' describes the process of "propagation and evaluation".

Element 4: This exclusion is based on, and draws principally from the language used in, TRIPs Article 27.2. The TRIPs Agreement does not, however, provide any additional guidance as to how this exclusion (or amalgam of exclusions) can or should be interpreted or implemented.

Some are sceptical about the ultimate utility and purpose of these exclusions, given that commercial exploitation of a product may continue independently of whether or not intellectual property protection is granted, as long as it does not infringe on other intellectual property holders' rights. They point out that it is inconceivable that a patent exclusion could be important to control genetically modified plants.

Others believe that the exclusion may be important to reduce incentives to develop such plants. Independent of the ultimate relationship between intellectual property protections and marketing success, they point out that intellectual property rights are granted by the state, and consequently, the conditions under which they are granted should be kept in line with the basic deep-rooted feelings and beliefs of society.

Some have suggested that *ordre public* might be violated by the insertion of characteristically[23] human genetic material into a plant variety. Others have suggested that all transgenic plant varieties are immoral.

Some feel that the test for morality should not be an abstract, universal concept, but rather a flexible one that is sensitive to the moral codes of local cultures. One suggestion was to include 'classes' of subject matter considered to violate public morality, such as those technologies that prevent farmers from using saved seed.

In any event, TRIPs allows members to decide for themselves what is contrary to *ordre public* or morality. Any innovation banned as a result certainly need not be protected by intellectual property law. Only perverse uses of such a ban could reasonably be challenged. (For discussion of the use of morality and *ordre public* in the context of disagreement over 'Terminator technology' and 'patenting life forms', see *Seeding Solutions*, Volume 1, pp 37–8 and 101–2 respectively.)

Element 5: While Article 1 addresses the general application of the provisions to plant species and genera, this element allows for a differentiation among different varieties of the same species or genus. In this way, this element reflects UPOV 1978, Article 2.2. A species or genus so treated still counts as one of the number required to meet the provisions of UPOV 1978 (see Article 2, Option 2, and the commentary). Furthermore, as noted in the commentary to Article 2, TRIPs does not explicitly mention the possibility of any exclusions of plant groupings. WTO members might therefore be obligated to protect all plant varieties of all species and genera.

Part Two: Definitions

Article 4: Plant variety

Option 1

A plant variety is that which is protectable pursuant to this Act.

Option 2

Plant variety means a plant grouping within a single botanical taxon of the lowest known rank. This grouping, irrespective of whether the conditions for the grant of rights conferred under this Law are fully met, can be defined by the expression of the characteristics resulting from a given genotype or combination of genotypes. It can be distinguished from any other plant grouping by the expression of at least one of the said characteristics, and considered as a unit because it is capable of being propagated unchanged.

Commentary

This article provides options for the definition of 'plant varieties'. It is not necessary to provide such a definition as long as the conditions for protection (which are considered separately in Part Three, below) are precise enough to define exactly what is protectable pursuant to the law.

Option 1: Following this option, the law would not include a definition of plant variety, other than a *de facto* reference to that which is protected pursuant to subsequent sections of the law concerning conditions of protection. In this way, the option is consistent with UPOV 1978.

Option 2: This definition is taken from Article 1 of UPOV 1991. The subject matter covered in this definition is broader than the class of plant varieties for which protection must be granted under UPOV 1991. The 1991 UPOV Convention first provides a definition of 'plant variety' and later defines a smaller subclass of protectable plant varieties, i.e., those that are novel, distinct, uniform and stable. The main purpose of this definition is to emphasize that the concept of 'variety' should not be assimilated to that of 'protected variety' or 'protectable variety' (see UPOV Model Law on the Protection of New Varieties of Plants, Geneva 1996, p 12).

It is interesting that the UPOV 1991 drafters decided to proceed by the method of defining 'varieties' first, and later defining the subcategory of them that could be protected, given that plant varieties as they are defined in the agreement are not part of the standard botanical taxonomy. While the International Code of Nomenclature does recognize '*varietas*', the definition is not the same as that in UPOV 1991 and in the national plant variety protection laws we are familiar with. The UPOV 1991 drafters could have simply skipped a step and defined plant varieties as those groupings of plants that satisfied the 'distinct, uniform, stable' (DUS) and novelty requirements. In this way, UPOV 1991 would have been more like UPOV 1978.

The fact that UPOV 1991 provides a broader definition of plant varieties than the class of plants for which that same agreement ultimately provides protection supports the argument that there is still room to use a looser criterion than that found in UPOV 1991 to extend coverage to a more heterogeneous plant grouping.

Article 5: Essentially derived varieties

Option 0

no provision

Option 1

Element 1

Essentially derived plant varieties are those

a) that are predominantly derived from another variety (the initial variety), or from a variety that is itself predominantly derived from the initial variety, while retaining the expression of the essential characteristics that result from the genotype or combination of genotypes of the initial variety;
b) that are clearly distinguishable from the initial variety, and, except for the differences which result from the act of derivation, conform to the initial variety in the expression of the essential characteristics that result from the genotype or combination of genotypes of the initial variety; and
c) that conform, except for the differences that result from the act of derivation, to the initial variety in the expression of the essential characteristics that result from the genotype or combination of genotypes of the initial variety.

Element 2

Essentially derived plant varieties may be obtained, for example, by the selection of a natural or induced mutant, or of a somaclonal variant, the selection of a variant individual from plants of the initial variety, back-crossing, or transformation by genetic engineering.

Commentary

The concept of essential derivation was introduced into UPOV 1991 to ensure that a variety that is essentially derived from a protected variety, and sufficiently distinct from that other variety, may be the subject of protection, but cannot be exploited without the authorization of the breeder of the protected variety. Elements 1 and 2 of Option 1 are taken from UPOV 1991 Article 14.5. However, if the initial variety is not protected, or protection rights over it have expired, no authorization is needed under UPOV 1991 for the exploitation of the essentially derived variety.

Essential derivation remains a contentious issue some ten years after its adoption into UPOV 1991. The difficulties stem from the principle known as the 'breeder's exemption' embodied in UPOV 1978, which states that "authorization by the breeder shall not be required either for the utilization of the [protected] variety as an initial source of variation for the purpose of creating other varieties or for the marketing of such varieties."

The breeder's exemption is considered essential for continued progress in plant breeding. Problems arose in its implementation, however. Some breeders charged that the exemption was being abused by other breeders, who, after making only slight variations to materials that the exemption allowed them to use, then claimed to have created new, protectable varieties. Through this practice — sometimes called 'copycat breeding'— it was felt that these breeders were taking unfair advantage of the former breeders' work, and reaping a disproportionate share of rewards as a result.

This problem is addressed by giving the original breeders the right to forbid exploitation of essentially derived varieties. Many argue that essential derivation provisions also have other important benefits: they are said to act as an incentive to preserve (or slow the erosion of) genetic diversity; to stimulate genuine varietal improvement; and to protect farmers and consumers against paying for insignificant adaptations.

Conversely, others maintain that the breeders' exemption should not be limited under any circumstances. Historic 'free access', the right of breeders to commercialize new varieties developed from protected varieties without having to pay any owners, has been, and still is, fundamental for the development of new varieties. Further, they charge that it is difficult, if not impossible, to prejudge the actual merits of seemingly minor varietal changes, that the valuation of 'essentially derived' will always remain subjective and subject to abuse and litigation. They argue that the buyer should be allowed to make this judgement.

The development of genetic engineering, opening new possibilities for piracy of varieties, helped speed up the revision of the Convention. The UPOV 1991 Convention introduced, with the full agreement of breeders' associations, the concept of essential derivation. As can be seen in Element 2, the concept of essential derivation plays an important role in establishing the relationship between plant breeders and biotechnologists. If a single gene is inserted in a protected plant variety, the derived variety cannot be commercialized without the prior authorization of the owner of the original variety.

Some have argued that essential derivation clauses in plant variety protection laws work to prevent copycat breeding of farmers' varieties. It is important to note, however, that the right to forbid exploitation of varieties that are essentially derived from previous varieties only applies in cases where the previous varieties are themselves the subject of plant variety protection. If a farmers' variety is not protected (and most of them are not), the inclusion of an essential derivation clause in a national plant variety protection law will do nothing to prevent exploitation of plants that are essentially derived from them.

Whether the rights conferred under an act protecting plant varieties should only extend to varieties derived from protected varieties, as is the case under UPOV 1991, or to varieties derived from unprotected varieties as well, is an issue we will address in Part Four, 'Rights conferred'.

We will also consider, in Part Three, conditions of protection that could have the effect of extending plant variety protection to a range of farmers' varieties that would not qualify for protection pursuant to UPOV 1978 and UPOV 1991 standards.

Article 6: Small-scale farmer

A small-scale farmer is a person who:

Element 1

sells [number]% or less of his or her annual harvest in a particular crop for food/commodities (conversely, a small-scale farmer personally consumes [number]% or more of his or her total annual crop harvest)

Element 2

cultivates [number] acres of land or less [with propagating material of the variety]

Element 3

harvests [number] tonnes or less of the protected variety

Element 4

produces less than [number] tonnes of agricultural crops

Element 5
farms an area less than that necessary to produce [number] tonnes of harvest

Commentary
The definition of small-scale farmer is important later, when exemptions from the rights conferred for small-scale farmers are considered (in Part Four). This article presents elements for a possible definition of 'small-scale farmers'; these elements are not mutually exclusive. For the purpose of transparency and simplicity, however, we recommend that a single element should form the core of the definition. In choosing limits, a careful balance must be struck between the interests of small-scale farmers, institutional breeders and society in general. Of course, striking such a balance is easier said than done, and we do not purport to have done so here.

Element 1: Defining 'small-scale farmer' by reference to the proportion of total yield used for personal consumption has the advantage of being very simple. Any farmer who is able to consume, say, 95% or more of a particular crop must not have a very large crop (or must consume an extraordinary amount of snow peas, barley or whatever he or she is growing.) It has been suggested that selling 5% or less is the proper balance. Determining the exact formula should be left to national policy-makers, as it will have to reflect each country's priorities and on-the-ground realities.

Element 2: Defining small-scale farmers by reference to the number of acres cultivated with the protected variety also has the advantage of being very simple. It should be noted, however, that the number of acres is only one indicator of the actual amount of the harvest. Element 2 would allow farmers to cultivate freely many protected varieties as long as the number of acres of each variety is limited. Commercial breeders may argue that the relevant area should be the total planted with protected varieties of any kind.

Element 3: By reference to a maximum number of harvested tonnes produced with the variety, this option addresses one of the drawbacks of Element 2. Under Element 3, farmers could still freely cultivate many protected varieties as long as the number of tonnes harvested is limited for each variety. Giving an absolute number of kilogrammes only makes sense if one has a well-defined social or geographic referent. Commercial breeders may argue that the relevant tonnage should be the total from protected varieties of any kind.

Element 4 refers to a maximum number of harvested tonnes of all agricultural crops produced by the farmer. Thus, independently of the amount of harvest obtained from a specific protected variety and of the number of acres used for each variety, farmers would benefit from any privileges granted to small-scale farmers, but only if their total harvest does not exceed a certain number of tonnes.

Element 5 reflects the model of the more complex Article 14 of Council Regulation (EC) No. 2100/94 of 27 July 1994 on Community Plant Variety Rights, which adopts the EU definition of 'small farmer' given in EU Council Regulation (EEC) No. 1765/92 of 30 June 1992.

Part Three: Conditions for the grant of protection

Part Three, Article 7 provides three different combinations of 'conditions of protection' for plant varieties. Policy-makers might opt to include only one of these combinations in their national plant variety protection law. On the other hand, it is possible to use two or all three combinations, with each applying to different species or groupings of varieties within species of plants. (For the latter reason, we have labelled the three different combinations 'elements' and not 'options'.)

Articles 8–11 consist of options for the definition of each of the conditions for protection included in Article 7.

Finally, in this part we consider two conditions of protection that have not yet been included in any domestic Plant Breeders' Rights laws: providing a certificate of origin (CO), and obtaining prior informed consent (PIC) from the suppliers of the genetic resources used by breeders in creating the varieties they seek to protect.

Article 7: Conditions of protection

Plant variety protection shall be granted for plant varieties that:

Element 1

a) have been invented, and are
b) new,
c) useful/capable of industrial application, and
d) non-obvious.

Element 2

a) are novel,
b) distinct,
c) uniform, and
d) stable.

Element 3

a) are novel,
b) distinct, and
c) identifiable.

Commentary

This article is one of the core provisions of the law. It states the requirements that a plant variety must fulfil in order to be eligible for protection under this law. The requirements (except 'invention', 'non-obviousness' and 'industrially applicability', which are discussed in Section Three, 'Options for intellectual property laws for biotechnological innovations') are defined and discussed in the following articles.

Element 1: The combination of conditions for protection set out here is identical with that required by TRIPs for the grant of patents. These conditions are more demanding (and harder to satisfy) than those set out in UPOV 1978 and UPOV 1991.

Element 2 is identical with the criteria set out in UPOV 1978 and UPOV 1991 (to be discussed in Articles 9 and 10).

Element 3: This element replaces the relatively strict requirements of uniformity and stability with the looser condition of 'distinctness and identifiability' (DI) (see Article 10 for a more detailed explanation). A DI protection requirement would not comply with the UPOV Conventions. This would not be a problem, of course, for countries that are not signatories to the UPOV Conventions. Despite not satisfying the UPOV standards, the criterion of identifiability may well satisfy TRIPs Article 27.3(b), which includes no obligation on WTO member countries to follow the UPOV model or to become members of UPOV. Being the widest in scope, Element 3 could be used as a national baseline criterion. Varieties that satisfy the stricter criteria could qualify for stronger and/or longer protection.

TRIPs 27.3(b) envisages patents, an effective *sui generis* system, or both to protect plant varieties. The United States currently offers both patents and Plant Breeders' Rights as forms of protection for plant varieties. Article 53(b) of the European Patent Convention (EPC) explicitly excludes plant varieties from patentability, and until recently, the European Patent Office (EPO) rejected patent applications for transgenic plants that embraced plant varieties in their claims. In December 1999, however, the Enlarged Board of Appeal of the EPO turned this practice around. Consequently, the EPO is now allowed to grant patents relating to plants whether or not these patents embrace plant varieties in their scope. Therefore, a plant variety may now be subject to a broad 'plant patent', which has the same effect on it that a narrow patent granted on that specific plant variety would have.

Whether the courts of the contracting parties of the EPO follow this precedent remains to be seen, but it is very likely, since the EU Directive on the legal protection of biotechnological inventions also allows patent claims embracing plant varieties. Although, pursuant to the Directive, plant varieties and animal races as such are, in principle, excluded from patentability, inventions involving plants or animals shall be patentable if the technical feasibility of the invention is not confined to a particular plant variety or animal race. More explicitly, the Directive states that plant groupings are patentable "*even if they comprise new varieties of plants*" [italics added]. Thus, under the patent laws of EU member states, patents may not be granted for a single plant variety or animal race, but may be granted for broader groupings and varieties or races.

Countries wishing to bring their legislation in line with TRIPs and yet limit the number of varieties protectable through intellectual property regimes may consider the option of not excluding plant varieties (or rather, excluding 'plants other than plant varieties') from patentability. By doing so, they need no longer introduce a *sui generis* system for plant varieties. Such an option is feasible because plant varieties will only rarely qualify as inventions, thereby satisfying the patent requirements. However, under these circumstances, transgenic plants would be the most likely candidates for patent protection, and breeders of conventionally bred plants would be severely disadvantaged.

Some members of the Crucible Group believe that the approach of granting only patent protection for plant varieties, in order to limit intellectual property protection for plant varieties, would be a major step backwards. While undoubtedly complying with TRIPs, this approach would take plant variety protection back half a century, arguably ignoring the important contributions that plant breeders can make, and would damage agriculture, commerce and industry in the country adopting it. This difference of opinion is encapsulated in Viewpoint box 16, 'Should patents on plant varieties be allowed?', in Topic 3, Section 3, Article 8, below.

Countries wishing to opt for Element 1 should refer to Topic 3, Section 3, Part 2 for options for the definitions of the requirements 'invention', 'non-obviousness' and 'industrial applicability'. The articles that follow discuss the conditions for the grant of protection to plant varieties and the rights conferred as per Elements 2 and 3.

Article 8: Novelty/grace period

Option 0

no provision

Option 1

The variety is novel if, at the date of filing of the application, or, where relevant, at the priority date, material of the variety has not been sold or otherwise disposed of to others, by or with the consent of the applicant or his or her successor in title, for purposes of exploitation of the variety:

a) in the territory of the country, for longer than one [or other number] year[s]; and

b) in a territory other than that of the country, for longer than four [or other number] year[s] or, in the case of trees or vines, for longer than six [or other number] year[s].

Option 2

The variety is novel if, at the date of filing of the application, material of the variety has not been sold or otherwise disposed of, by or with the consent of the applicant or his or her successor in title, for purposes of exploitation of the variety:

a) for longer than one [or other number] year[s];
b) in the case of [name species/genera], for longer than six [or other number] year[s]; or
c) in the case of varieties which have been bred by small-scale farmers [or other groups] and only cultivated within limited areas of the country, for longer than ten [or potentially much larger number] year[s].

Commentary

Novelty, as understood here, is considered with respect to any possible use made of the variety *with the consent of the breeder* before the application for protection (e.g., marketing of the variety for up to one year). Novelty as defined here relates to what is known as a 'grace period' in the context of patent law, *and differs substantially from the meaning of 'novelty' in patent law*. 'Grace period' means a period within which an inventor may have sold, offered for sale, publicly used or published an invention without its novelty being compromised or destroyed.

Novelty may pose a significant hurdle for the protection of traditional varieties that have been sold or disposed of to others for many years with the consent of their creators and, therefore, could not be considered to be new under current Plant Breeders' Rights legislation. This problem may be overcome by extending the grace period (see Option 2). On the other hand, it should be noted that many of the so-called traditional varieties undergo changes due to the continued breeding efforts of rural communities. If such varieties have been disposed of to others or sold for the purpose of exploitation for only a limited period, novelty should not pose an insurmountable obstacle to their protection.

Detailed exemptions related to novelty can be found in both UPOV Conventions. These exemptions include, for example, cases where propagating material of a plant variety has been disposed of to third parties for small-scale processing trials or to a statutory authority for the purpose of biological security testing. These exemptions are not considered in detail in this article.

Option 0: From a purely legal perspective, there is no need to grant a grace period for a variety. However, it is desirable from a breeder's point of view. Most countries' patent laws do not grant any grace period (during which an invention may be disclosed to the public without loss of rights) for filing an application. This means that, except where priority of a previous filing in another country may be claimed (see discussion below), any disclosure that allows someone to reproduce the invention eliminates the possibility of getting a patent in these countries.

Option 1 draws mainly from UPOV 1978 and UPOV 1991 and differentiates between domestic and foreign selling or disposal of the protected varieties. Reference is made to a so-called priority date. The right of priority is relevant only in the international context and usually requires the establishment of an international agreement (such as the Paris Convention for the Protection of Industrial Property). The right of priority means that on the basis of the first filing of an application for protection in one of the contracting states (of the Paris Convention, for example), the applicant may, within a certain period of time, apply for protection in any of the other contracting states. These later applications will then be regarded as if they had been filed on the same day as the first application.

Under TRIPs, WTO members have no obligation to grant a right of priority in their *sui generis* system for the protection of plant varieties. This is because the right of priority (as referred to in TRIPs 2.1) is limited to patents (and utility models, where they exist), industrial designs and trade marks.

Option 2 modifies Option 1 with regard to three major components of the definition of novelty:

1. Option 2 does not differentiate between domestic and external sale/disposal of the variety. Consequently, countries cannot grant a grace period solely for varieties that have only been sold domestically or that have only been sold externally. The economic implications of this approach should be examined carefully;
2. Option 2(b) would allow the list of species and genera to be extended beyond trees and grapevines, for which a longer grace period shall be applied (as in the UPOV Conventions). This would allow more flexible adaptation of the grace period to the actual needs of the breeding sector and the seeds market in general; and finally,
3. Option 2 provides a possible exemption for varieties that have been bred by small-scale farmers (or other groups) and that have been cultivated only in limited areas of the country. This option would have the effect of extending the grace period for farmers' varieties.

Recommendation

The Crucible Group recommends that any country introducing a plant variety protection law should grant a grace period during which a variety may be sold, offered for sale, publicly used or tested without its novelty being destroyed or compromised in any way.

The last two components — the extended list of species and the exemption for traditional varieties — could also be combined with the grace period provision included in Option 1.

Article 9: Distinctness

The plant variety is distinct if:

Option 1

it is clearly distinguishable from any other variety whose existence is a matter of common knowledge at the time of the filing of the application [or, where relevant, at the priority date].

Option 2

it is clearly distinguishable by one or more characteristics of agronomic or other practical relevance [or with respect to a given distribution of such characteristics] from any other variety whose existence is a matter of common knowledge at the time of the filing of the application [or, where relevant, at the priority date].

Common knowledge of another variety is established, in particular, if that variety has been entered in an official register of varieties or such entry has been requested; if it has been precisely described in a publication or included in a reference collection; or if it is being cultivated or marketed.

Commentary

Distinctness in this article relates to 'novelty' in patent law (not to be confused with novelty and grace period in Article 8) and refers to the relationship between the variety for which protection is sought and any other existing variety. In other words, the requirement of 'distinctness' aims at preventing varieties that are 'common knowledge', or as patent lawyers say, that form part of the 'state of the art', from being protected.

Option 1 draws mainly on UPOV 1978 and UPOV 1991. A fundamental feature of the UPOV Convention, now embodied in Article 12 of UPOV 1991, is that protection shall only be granted after an examination to determine if the variety is novel and clearly distinguishable from all other varieties that are a matter of common knowledge. The 1978 Convention did not define 'common knowledge', but provided a non-exhaustive list of examples of how a variety could become a matter of common knowledge. When the Convention was revised in 1991, it was noted that the list of examples included events that would not necessarily be known to the public, for example, the addition of a variety to a reference collection. Accordingly, UPOV 1991 leaves 'common knowledge' undefined and specifies only that certain acts (not likely to be known to the general public) shall be deemed to render varieties a matter of common knowledge; it is, therefore, subject to 'common-sense' understanding or interpretation. A variety that is a candidate for protection must be clearly distinguishable from any variety that is a matter of common knowledge anywhere in the world. Under both Conventions the variety, in order to be distinct, does not have to confer an additional value to the variety.

Option 2 modifies Option 1 with regard to the importance of characteristics relevant for the distinctness test.

Article 10: Uniformity, stability, identifiability

Option 1

The variety is uniform if, subject to the variation that may be expected from the particular features of its sexual reproduction or vegetative propagation, it is sufficiently uniform in relevant characteristics.

The variety is stable if its relevant characteristics remain unchanged after repeated propagation or, in the case of a particular cycle of propagation, at the end of each such cycle.

Option 2

The variety is identifiable if [with respect to the characteristics of its plants or with respect to a given distribution of characteristics among plants] it can be identified by a person skilled in the art.

Commentary

Option 1: Under the UPOV Conventions, a variety has to be "sufficiently homogeneous" (UPOV 1978) or "sufficiently uniform in its relevant characteristics" (UPOV 1991), subject to the variation that may be expected from the particular features of its propagation. The UPOV requirement on distinctness attaches great importance to uniformity standards. *The uniformity of a variety must be established for the decision on distinctness*, i.e., only those characteristics in which both the candidate variety (for protection) as well as the similar varieties are uniform can be used. Different degrees of uniformity are not accepted as characteristics determining distinctness.

To be considered homogeneous according to the existing UPOV Test Guidelines,[24] the variation shown by a variety, depending on the breeding system of that variety, must generally "be as limited as necessary to permit accurate description and assessment of distinctness and to ensure stability". No doubt, this definition implies a certain tolerance depending on the different reproductive systems of varieties — a cross-pollinated variety must be judged in a different way than a vegetatively propagated one. Whereas the maximum acceptable number of off-types is defined exactly for vegetatively propagated varieties and self-pollinated varieties, tolerance limits in cross-pollinated varieties are set up only through comparison with comparable varieties already known.

Careful attention must be paid to stability when testing for distinctness and uniformity. According to UPOV, a variety is deemed stable if its relevant characteristics remain unchanged after repeated propagation or, in the case of a particular cycle of propagation, at the end of each such cycle. The relevant characteristics are either those used for distinctness or those included in the variety description drawn up on the date the protection was granted. If the variety is not stable, it will no longer be the same variety but a different one, as the relevant characteristics (those listed in the variety description drawn up on the date the protection was granted) will have changed.

The requirement of homogeneity or uniformity has been the subject of criticism by those concerned about the erosion of agricultural genetic diversity. They contend that by rewarding only the breeding of uniform varieties, today's Plant Breeders' Rights laws create perverse incentives to 'breed-out' diversity. They argue that this is the opposite of the desired effect, particularly in high-risk subsistence farming situations where a higher degree of genetic variability in crop plants is an important aspect of food security. Consequently, they argue that the uniformity/homogeneity requirement cannot be justified on either agronomic or practical grounds.

In response to such allegations, some members of the formal-sector plant breeding community argue that even a term as central to these debates as 'biodiversity' is used and understood differently in different disciplines, thereby weakening the basis of allegations turning on a presumed common appreciation of such terms. They claim that lack of understanding of breeding techniques and concepts such as variability in a population on the field and at a landscape level has led to overestimates of the impact of new variety development on biodiversity and the impact of Plant Breeders' Rights on the development of new varieties.

Option 2: One of the purposes outlined in Article 1 is to create intellectual property protections for farmers' varieties that are not protected by existing Plant Breeders' Rights laws.

One way to approach the creation of such protection would be to replace the DUS requirements for protection with that of identifiability (DI). 'Distinctness' defines a variety in terms of all its morphological characteristics. The term 'identifiability' emphasizes the legal need to identify the protected subject matter rather than to characterize it fully. A typical combination of a few characteristics may in many cases suffice to define a right. This term leaves considerable and explicit flexibility for interpretation, and it underlines that uniformity is not a goal in itself, but that identifiability remains necessary.

Crucible Group members are divided over whether or not these criteria are desirable or practically implementable. The parameters of the Group's debate are set out in Viewpoint box 12.

Viewpoint box 12: What conditions for protection are most appropriate in a plant variety protection law?

The three technical criteria that are most often taken into account at the moment are distinctness, uniformity and stability (DUS). Some members of the Crucible Group feel strongly that those criteria have been proven efficient and that it would be a mistake to drop them in any law on protection of plant varieties. Others consider that the criteria of uniformity and stability currently required by national authorities and international guidelines set much higher standards than are justified, creating incentives for unnecessary and sometimes dangerous uniformity and preventing farmers' varieties from being protected.

Keep DUS.

1 The uniformity criterion is very flexible and takes into account the particular features of propagation of the variety. In addition, the UPOV wording is 'sufficiently uniform'. That should allow protection of farmers' varieties.

2 The allegedly over-narrow limits on heterogeneity in many UPOV-compliant systems are not due to the DUS criteria. They result from test guidelines set up by UPOV and national authorities as a means of implementing those criteria. More heterogeneity could be recognized simply by altering these guidelines. There is no need to change the DUS criteria themselves.

3 There has been no scientific evidence of any decrease in genetic diversity measured by coefficients of co-ancestry and molecular markers in a given crop since the development of so-called modern varieties fulfilling DUS criteria.

4 Modern varieties, when they have been selected in a given environment, are at least as stable and often more stable than local cultivars. They are also, even in areas of low productivity, better yielding.

5 If the variety to which a title is granted were not stable, then the protection would be useless. If after four or five years the variety has changed, the protection is lost, and it is misleading to advise governments to protect unstable varieties.

6 If 'identifiable' varieties are protected, and if they shift in the same direction, who will be the owner of the rights over the 'evolved' varieties? This would lead to endless disputes.

DI should be applied.

1 The identifiability criterion provides maximum flexibility by pointing to a strictly legal need rather than mixing the physical properties of the plants with the legal needs of identification.

2 The concept of identifiability can be well applied by persons skilled in the art, which may often include local farmers, and it seems particularly suited for many orphan crops, 'poor peoples' crops', and traditional and recent 'farmers' varieties', which often would not withstand the current tests for 'uniformity'.

3 Stability becomes an unnecessary criterion, because it is clear that if the variety becomes unidentifiable in later generations, it does not fall under the scope of the rights.

4 The problem of 'shifting and drifting' varieties is well known for (protected and unprotected) open-pollinated varieties already. How 'broad' a variety can be will always be an issue that must be judged, whether by uniformity and stability or by identifiability. This is particularly true regarding essentially derived varieties, where it is necessary to determine whether a new, different variety has been created.

5 The concept of identifiability opens many new possibilities to make

7 DUS may facilitate membership in UPOV, which requires these criteria.

8 How is scope to be determined? 'Identifiability' is the criterion used in patents, and the scope is determined by claims drafted by the applicant. Many difficulties have arisen in biological patents when such claims have been drafted too broadly. Is it proposed to have 'claims' in plant variety rights applications? Will these be examined by the receiving office for undue breadth? What criteria will be applied? This loses the advantage of UPOV systems, in which legal representation to obtain rights is usually unnecessary. 'Identifiability' will make for expense and uncertainty.

more heterogeneous populations protectable, as long as they share some specific properties that make them different from other populations.

6 Clearly, the concept of identifiability is also applicable to all varieties that have hitherto been termed 'uniform and stable'. So, while national authorities may become much more flexible, they do not have to put such varieties at any disadvantage.

Article 11: Public-benefit requirement

Option 0
no provision

Option 1
In addition to the requirements mentioned in the previous articles, the rights shall be granted only if the variety has a higher value for cultivation and use than any other protected or otherwise available variety.

Option 2
In addition to the requirements mentioned in the previous articles, the rights shall be granted only if the variety meets the needs of particular agricultural environments and national priorities. The government may pass regulations setting out these priorities regarding plant variety development, taking into consideration the following factors:

Element 1
the need to encourage the planting of several different varieties of the same species within fixed geographical areas

Element 2
the need to encourage the breeding of plant varieties with high degrees of genetic variability

Element 3
the need to develop varieties that respond well to local, specific environmental conditions

Element 4
the need to develop varieties that will perform well in marginal agricultural lands

Element 5
the need to develop varieties that meet different social, economic or cultural needs

Commentary

To the extent that these additional requirements can be construed as creating extra, stricter criteria for protection, they would not comply with the UPOV Conventions.

There is considerable controversy regarding the effect of intellectual property protection for plant varieties on, among other things, crop biological diversity and expenditures for research and development designed to meet local farmers' needs. National governments considering such legislation will need to undertake further studies to ensure that they understand the ultimate effects of such policy manipulations on the environment.

Option 0: This option has been adopted in most current UPOV systems. It is preferred by industry since it gives rise to less uncertainty about whether protection will be granted.

Option 1 refers to the so-called value in cultivation and use (VCU) requirement.

This option has some precedent in many countries' seed registration laws. Some countries do not allow plant varieties to be sold for agricultural use unless they are first 'registered' in an official list. This registration is, in principle, quite independent of whether such varieties are proprietary. Before allowing varieties to be added to the list, agricultural authorities require evidence of uniformity and stability, and, often, of VCU. Varieties that do not show useful properties (such as increased yield or resistance to diseases) compared with varieties already on the list are not added to it. Although countries like Germany originally had the VCU requirement in their plant variety protection legislation, it seems that no country presently applies VCU as a condition of plant variety protection laws.

Option 2: In passing regulations such as those envisaged in Elements 1–5, the government would have to be sensitive to the position of small-scale breeders and not enforce conditions too strictly when such businesses simply could not afford to comply.

Option 2, Element 2 requires "high degrees of genetic variability". Some members of the Crucible Group point out that it could be difficult to define

Viewpoint box 13: Should public benefit or value in cultivation and use (VCU) be included as conditions of protection?

Some members of the Crucible Group consider that a public-benefit requirement should be added to the protection requirements discussed above. Others consider such criteria to be difficult to implement in the frame of an intellectual property protection law. The following arguments have been offered for and against VCU.

No.

1 Intellectual property protection rights are valid within a national territory. Certain varieties may offer significant advantages in some parts of a country, but not in others. Likewise, a variety with, say, a given disease resistance may yield less than a non-resistant variety when the disease is absent. But when the incidence of the disease is high, the value of that variety increases dramatically. How are such differences to be reconciled in the context of VCU? Is it at all possible to reconcile such differences within the context of national legislation? If so, how would it be implemented?

2 Varieties that offer an advantage only under certain environments or conditions should be protected. Who better than farmers to attest the value of a variety?

3 Past efforts by governments to limit farmers' choices (e.g., by financing only particular approved varieties) have not worked well.

4 Intellectual property laws have been implemented in several countries for more than a century. Almost nowhere, for the reasons mentioned above, has the criterion of value been introduced.[25] (Here it is important to highlight that the criterion of 'value' is distinct from that of 'utility', where the invention can be used industrially without judging the value of the invention.)

5 Despite the arguments above, if some nations feel that the VCU characteristics of varieties grown within their boundaries must be regulated, this can be done through means other than protection of intellectual property. National catalogues or registers are used in many countries.

Yes.

1 VCU can be used to better focus the incentives provided by Plant Breeders' Rights. Instead of giving non-specific incentives for all kinds of innovations and then trying to regulate market access in a second stage (as many countries do today), it may be more effective to set targets and standards for innovations aimed at meeting national agricultural priorities as determined by the appropriate authorities.

2 The notion of usefulness is integrated into many different intellectual property laws, and there is no reason why it should not be taken to mean 'social usefulness' as judged by legitimate democratic bodies. Some countries, such as Germany, had VCU as a criterion in their Plant Breeders' Rights laws. It was dropped only in view of international obligations of reciprocity. Such international obligations do not exist for most countries today, and the related legal problems can be overcome by specific clauses, for instance, to ensure that potential changes in VCU status do not affect novelty in third countries.

3 Historical experience shows that it is entirely possible to regulate VCU on a national level by taking differences in regional conditions into account. Many European countries have done so and continue to do so, albeit not by linking this process to the granting of intellectual property rights in plant varieties. The link with rights, however, can be made, and VCU can be used as an instrument of agricultural-policy agenda setting.

and implement this condition. Phenotypic and physiological characteristics commonly used to describe varieties constitute only a fraction of the variability that exists within plants of a species. The use of molecular information is relatively new, and there is little agreement even in the scientific community on what constitutes a satisfactory level of variability or genetic 'distance' within varieties.

Article 12: Certificate of origin

Option 0

no provision [i.e., no certificate of origin is required from patent applicants]

Option 1

In addition to the requirements mentioned above, the rights shall be granted only if the applicant has presented a certificate of origin (CO).

A CO must specify the country [and the community] of origin of the varieties from which the subject variety is derived.

Commentary

In the post-CBD world, it is readily accepted that those moving germplasm out of a country must have the sovereign state's permission to do so. Many national access laws already require such permission, from both the national government and the community from which parties are obtaining biological resources. A provision in a national plant variety protection law requiring a CO would be consistent with both the CBD and these national access laws. Despite this consistency, to date there is no international agreement requiring a CO as a condition for protection for plant varieties. (Denmark tried but failed to have something like a CO obligation included in the European Patent Directive.) Nor are there any national plant variety protection laws that require a CO as a condition of protection.

Nevertheless, proponents of the idea consider a CO an easily implemented criterion for protection. They believe that resistance to the concept stems from fear that the process of obtaining COs will raise expectations of shared benefits, and that once enough COs were being issued, advocates for traditional farmers would be able to make a stronger renewed bid for a PIC requirement (see Article 13). Opponents of COs say that they foresee many difficulties, both theoretical and practical, in implementing such a requirement.

Article 2 of the CBD defines 'country of origin' for germplasm as "the country which possesses those genetic resources in *in situ* conditions". This implies that the country of origin is where domesticated or cultivated genetic resources developed their distinctive properties. However, it might technically be very difficult, if not impossible, to determine where some species developed their distinctive properties. The identification of the country of origin for the purpose of these provisions may also raise additional problems where any of several countries could be the country of origin and it is unclear where the material used actually originates. There is also some concern that the establishment of the 'current-time' CO could jeopardize the possibility of other countries or communities affirming their historic relationship to the material in question. Another area of concern is the achievement of intercommunity consensus in the country of origin or intergovernmental consensus in countries of origin in cases where the germplasm has spread over a wide area.

Questions arise about the procedures that would be used in connection with the CO requirement. Sceptics charge that it would not be sufficient simply to have breeders provide the certificates based on their own accounts of where material is from. They argue instead that the CO should come from the country of origin of the material. In either case, how will the necessary information be verified (for example, the parentage of his or her variety as documented by the breeder)? What is the penalty for failing (or being unable) to provide the information needed for a CO? What could applicants for protection do if the relevant authorities of the country of origin were slow in responding, or completely ignored requests for certificates? Would mistakes in documentation be correctable? The most obvious practical consequence of failure to obtain a CO would be that a breeder would not qualify for a grant of rights pursuant to these provisions. Another alternative could be that applicants who fail to provide a CO would not be prohibited from obtaining protection, but they would be subject to a requirement to reapply every year until they supplied a CO. Fines could be levied (or significant reapplication fees required) on each occasion that the patentee failed to supply the CO.

A country requiring a CO as a condition for granting protection must give some thought to the implementation of such a law in light of the difficulties discussed above. International coordination may be required for those species whose range spans more than one national boundary.

Article 13: Prior informed consent

Option 0

no provision [i.e., applicants do not need to demonstrate that they obtained the prior informed consent of the suppliers of genetic material that they used in creating their plant varieties]

Option 1

In addition to the requirements mentioned above, the rights shall be granted only if the applicant obtains the prior informed consent of the country of origin [and/or] the community of origin of the material used in breeding the new variety.

Commentary

In this law the two concepts, PIC and CO, have been split, since they are not necessarily linked, i.e., the breeder could be obliged to provide a CO but not be obliged to show that he or she obtained PIC. A CO is a less onerous requirement than PIC.

It should further be noted that the PIC requirement might conflict with the breeders' exemption inasmuch as PIC requires authorization for activities that are unrestricted under the breeders' exemption. (For a more detailed discussion of this issue, see 'PIC and the breeders' exemption' in the Appendix.)

Some people hold that the obligation to (a) trace ancestry, and (b) obtain the consent of communities who can lay some claim to genetic ancestors to new varieties would 'freeze' the movement of breeding material, rendering new plant breeding impossible.

Some seed companies, for instance, may do 300–400 crosses per year. Each cross might require PIC from nine or ten communities. In theory, this could oblige a company to obtain up to an impractical 4000 PICs per year. Moreover, how far back should the ancestry of the plant be traced for the purposes of obtaining PIC? Should it be limited to the last three generations of the new variety, or should one only need to get PIC for varieties or lines developed within a predefined number of years? Both these actions would exclude any previous generations of breeders in the plant's ancestry from obtaining any benefits. What if ancestry were established, and more than one community and/or country were identified, but all parties did not give their consent? Should the PIC requirement be limited to apply only to those ancestral varieties that came into existence after the requirement was passed into law? Of course, this would be highly controversial, for even as the proposed legislation would extend some control and benefit-sharing guarantees to traditional breeders, it would simultaneously extinguish claims to benefits derived from the entire history of plant breeding up to the date this legislation came into force.

Another issue regarding PIC concerns the technological feasibility of (a) establishing a connection between genetic ancestry and a particular community, and (b) determining the genetic ancestry of a variety under examination. Consider the case of material collected in the past and held in genebanks. The geographic origin of the material is often listed as the name of the country from which the material was received and not the country where the germplasm may have originated, nor the community, if any, from whom it was obtained. Determining the ancestry or pedigree of a variety is, relatively speaking, easier where breeding records exist. For instance, pedigree data for over 600 000 lines of rice are available in the International Rice Information System at the International Rice Research Institute (IRRI) and data for over 2.4 million lines of wheat are available in the International Wheat Information System at the International Maize and Wheat Improvement Center (CIMMYT).

The more interesting question of whether a particular variety of unknown provenance has any genes from, say, Thailand in it is obviously more complicated. Molecular techniques can help with identification and ancestry to some extent. If molecular markers unique to KDML (a Thai rice) are available and these are found in the variety being examined, the answer is 'yes'. The task could be likened to a phylogenetic problem where markers are run on test lines and on all putative ancestors, genetic distances computed and a tree constructed. One could probably tell which were ancestors and which not, although it is not certain whether that would stand the test of reasonable doubt. Such a method, however, is not feasible on a large scale. One solution in the future could be to introduce DNA identification tags (in the form of markers) into breeding lines for later identification. But would farmers be able to use such sophisticated techniques in their innovations?

The commentary clearly shows that the difficulty in implementing PIC lies in the details. One may certainly shape the PIC requirement in such a way as to cope with most of the concerns and difficulties mentioned.

Part Four: Rights conferred

Introduction

This part addresses the question of what rights rights-holders actually have if the variety fulfils all the requirements set out in Part Three above. Article 14 defines the parties in whom the rights created by the law will vest. Articles 15 and 16 should be read in combination. While Article 15 defines the acts which plant breeders could have the exclusive right to perform, Article 16 defines the variety of materials to which these acts could relate. Finally, Article 17 lists possible exemptions from the rights conferred. Article 17, therefore, should be read in the light of Article 15's definitions of the acts, since if an act is not part of the rights conferred under this law, it would make no sense to include any exemptions regarding that act.

Article 14: Entitlement to the rights

Element 1

Entitlement to the rights conferred under this Law shall vest in the person who [bred] [discovered and developed] the variety or in his or her successor in title.

Element 2

Where two or more persons have [bred] [discovered and developed] a variety jointly, entitlement to protection shall vest in them jointly. Subject to any agreement to the contrary between or among the joint breeders, their shares shall be equal.

Element 3

Where the person who [bred] [discovered and developed] is an employee, entitlement to the rights conferred under this Law shall be determined by the employment relationship in the context of which the variety was [bred] [discovered and developed] and in accordance with the law applicable to that relationship.

Element 4

Where an application has been filed by a person who is not entitled, persons entitled may submit a request to the competent authority for assignment of the application. Where the rights have already been granted to the person not entitled, the person entitled may submit a request for transfer of the title.

Commentary

These provisions address the entitlement to the exclusive rights conferred under this act. The legitimate rights-holder could be an individual breeder, a group or community of breeders, or a breeder employed by a breeding company.

Elements 1–3 address three typical situations under which a plant variety may be developed. All the elements include (in brackets) different activities a breeder must perform in order to qualify as a rights-holder. Concerning those activities, the first question is whether Breeders' Rights should be granted only if some intentional activity has been performed with regard to the variety. If this is the case, one would still have to decide whether collecting or discovering a variety should suffice to qualify the collector or discoverer as a rights-holder, or at least whether some breeding or developing must have been performed. To clarify the level of activity required, one could define in more detail the term 'plant breeding'.

Element 4: This provision addresses the case where a person or people, who under the previous elements are not entitled to the rights, have applied for or even been granted the rights. In these cases, the person or people entitled may request assignment of the application or transfer of the title.

Article 15: Acts requiring the rights-holder's authorization

Option 1

The following acts in respect of material of 'varieties' (defined in Article 16, below) shall require the rights-holder's authorization:

Element 1

producing or reproducing (multiplication) for commercial marketing

Element 2
producing or reproducing (multiplication)

Element 3
offering for sale [for the purpose of propagation]

Element 4
selling [for the purpose of propagation]

Element 5
marketing [for the purpose of propagation]

Element 6
conditioning [for the purpose of propagation]s

Element 7
exporting [for the purpose of propagation]

Element 8
importing [for the purpose of propagation]

Element 9
stocking for the purpose of the acts defined in Elements 2–8

Element 10
selling or advertising the protected material by referring to the plant variety protection seal

Element 11
using the variety for any purpose without adequate moral recognition of the rights-holder as the breeder of the plant variety

Commentary

The idea here is to choose activities that would be subject to the exclusive right of the plant breeder. For example, the rights-holder could be granted the exclusive right to import material of the protected plant variety into the country. At the same time, the rights-holder might not be granted the exclusive right to export the protected variety to another country. In such a case, one would still have to get the permission of the plant breeder to grow the plant, but the plant breeder would not be able to prevent third parties from exporting the material.

One principle worth considering is that the kind of rights conferred should be related to how broad or narrow the range of protectable subject matter is. For example, if a plant variety protection act were to recognize all 'identifiable' varieties, it would extend to a wide range of plant varieties. Consequently, one might wish to limit the rights one grants in association with 'identifiable' varieties. Otherwise, it is possible to foresee a situation where a very broad range of plants would be subject to very strong rights, and little could be done with plant material without the consent of the plant breeder.

One way of dealing with this situation would be to limit the rights associated with 'identifiable' plant varieties to the right of attribution and the sole right to sell the variety under a government-sanctioned and specific seal. Conversely, for narrower categories of plant varieties (i.e., those that are DUS), one might grant more substantial rights, such as the exclusive right to grant or refuse permission for reproduction. On the other hand, there may be very compelling reasons for extending equally strong (or stronger) protection to 'identifiable' varieties, so that incentives are created for small-scale farmers to continue innovating. Alternatively, it may be considered necessary to extend strong rights over identifiable varieties to indigenous and local communities because the protection of indigenous and local knowledge is of paramount concern.

Elements 1, 3 and 5 reflect the scope of rights as required by UPOV 1978. Elements 2–9 reflect the scope of rights as required by UPOV 1991. See commentary to Article 16, Elements 1–3.

Elements 2–9: see commentary to Article 16, Elements 1, 2, 4 and 5.

Element 10: Governments may choose to grant a plant variety protection seal to the breeder. The grant of a seal for a protected variety would be arguably in harmony with TRIPs Article 27.3(b), but would not comply with the UPOV Conventions.

If such a seal system exists, then this element comes close to reiterating trade mark law, as the seal would function like a trade mark. Under trade mark law, when a clear association between a name for the product and the product provider (here the breeder) can be demonstrated, the provider can apply to have the exclusive right to market the product under that name. If one were to follow up on the implementation of this option in this way — and do little more than create an exclusive right to market under a particular name — it could be argued that this would be a simple application of trade mark law. In that case, this option would be redundant and, therefore, would not fulfil the TRIPs requirements. However, if the application of the right to market exclusively under a registered name were limited to those situations where the breeder had satisfied a governmental body as to specific traits and/or qualities of the variety, an element missing in trade mark law would be added, and this option would represent a significant new *sui generis* element in the law. The use of a variety's denomination remains untouched by the seal.

Element 11: This is the same as a right to attribution. To be meaningful, this right should continue even after the original plant breeder has alienated (sold) his or her interests in the variety. The idea that an originator/breeder might have rights in his or her creation even after it has been sold is embraced in the intellectual property doctrine of 'moral rights' (which are often framed as separate from an originator's economic rights). Issues regarding the strength and content of moral rights usually arise in the copyright law context (and not the Plant Breeders' Rights or patent law context). An example of the principle in copyright is that a book author continues to be entitled to be recognized as the author of that book even if other rights to the book have been sold (Article 6 bis, Berne Convention for the Protection of Literary and Artistic Works). Another example is that a painter continues to have the moral right to have the 'integrity' of his or her painting preserved no matter how many times the painting has been sold. A painter has a right to legal action against owners down the line who might deface or manipulate the painting in order to use it for advertising or for the purposes of incorporating it in larger works of art — even though he or she no longer owns it.

Article 16: Material

Option 1
Material is:

Element 1
vegetative propagating material

Element 2
reproductive propagating material

Element 3
ornamental plants or parts thereof when used commercially as propagating material in the production of ornamental plants or cut flowers

Element 4
harvested material, including entire plants and parts of plants, provided this has been obtained through the unauthorized use of propagating material of the protected variety and the breeder has had no reasonable opportunity to exercise his or her rights in relation to the propagating material

Element 5
the materials mentioned in Elements [1], [2], [3], [4] from varieties essentially derived from the protected variety [from varieties which are not clearly distinguishable from the protected variety, or from varieties whose production requires the repeated use of the protected variety]

Element 6
products made directly from harvested material derived from material mentioned in Elements [1], [2], [3], [4], [5], provided that this material has been obtained through unauthorized use and that the breeder has had no reasonable opportunity to exercise his or her rights in relation to it

Commentary

This provision should be assessed in the light of the acts defined in the previous article. Countries are free to mix and match the different alternatives. Under TRIPs, WTO members are free to choose from this list. There is no obligation to accept the whole package of materials. It would be worthwhile to investigate the implications of the inclusion or exclusion of each of these various elements from the list.

Elements 1–3: According to UPOV 1978, reproductive and vegetative propagating material shall be subject to the exclusive rights mentioned in Elements 1, 3 and 5 of Article 15. Member states can, however, grant more extensive rights in respect of certain botanical genera or species, including the possibility to extend the rights to the marketed product.

Elements 1, 2, 4 and 5: According to UPOV 1991, the material mentioned in these elements shall be subjected to the exclusive rights mentioned in Article 15, Elements 2–9. However, acts in respect of harvested material (Element 4) will require the rights-holder's authorization only in cases where the harvest has been obtained through the unauthorized use of protected material and the breeder has had no opportunity to exercise his or her rights in relation to that material. Whether acts in relation to products made directly from such harvested material (see Element 6) will require the rights-holder's authorization depends on the decision of each contracting party.

Some Crucible Group members strongly believe that the extension to 'harvested material' obtained through unauthorized use of seeds, and especially to products made directly therefrom, is a potentially slippery slope towards grossly increased control by rights-holders over 'downstream' products and uses of harvested material. These members, therefore, advocate that governments should exercise caution in extending protection so far. Other members think that the extension to harvested material would be risky only if protection would extend to harvest that has been obtained through the authorized use of the protected material. They argue that the extension to 'harvested material' is absolutely necessary given the globalization of the commodity market.

Article 17: Exemptions from the rights conferred

Paragraph 1: Non-commercial use

Option 0

no provision

Option 1

The rights conferred on rights-holders in Article 15 shall not apply to activities done privately and for non-commercial purposes.

Paragraph 2: Research

Option 0

no provision

Option 1

The rights conferred on rights-holders in Article 15 shall not apply to experimental activities.

Paragraph 3: Farming

Option 0

no provision

Option 1

The responsible minister may restrict Breeders' Rights in relation to varieties protected by these provisions in order to permit [small-scale] farmers to plant-back seed which they have harvested on their own holdings.

Option 2

The responsible minister may restrict Breeders' Rights in relation to varieties protected by these provisions in order to permit [small-scale] farmers to plant-back seed which they have harvested on their own holdings [see Option 1] and to exchange that seed with other farmers on a non-commercial basis.

Option 3

The responsible minister may restrict Breeders' Rights in relation to varieties protected by these provisions in order to permit [small-scale] farmers to plant-back seed which they have harvested on their own holdings [see Option 1], to exchange that seed with other farmers on a non-commercial basis [see Option 2] and to sell seeds in the following limited quantities [the law would provide a list of maximum quantities in relation to the planting-back, exchanging and selling].

Paragraph 4: Breeding

Option 1

Authorization by the holder of the rights conferred in Article 15 shall not be required for the use of the variety protected by those rights as an initial source of variation for the purpose of creating other varieties, or for the marketing of such other varieties. Authorization shall, however, be required when the repeated use of the variety protected by rights is necessary for the commercial production of another variety.

Option 2

The rights conferred on rights-holders in Article 15 shall not apply to acts done for the purpose of breeding other varieties and shall not apply to any acts in respect of such other varieties, provided these other varieties have not been essentially derived from a protected variety.

Paragraph 5: Local origin

Option 0

no provision

Option 1

Residents of [the implementing country] shall be allowed to use plant varieties derived from or based on germplasm collected in this country without seeking permission from the rights-holder [on payment of royalty at a rate to be determined by government regulation].

Paragraph 6: Customary use

Option 0

no provision

Option 1

The rights conferred on rights-holders in Article 15 shall not apply to the customary uses of protected varieties by:

Element 1
small-scale farmers

Element 2
local communities

Element 3
indigenous communities

Commentary

This article addresses possible exemptions. Some of these exemptions apply only if the relevant acts are actually covered by the rights conferred on the rights-holder (Article 15).

Paragraph 1 (Non-commercial use): If, as required under UPOV 1991, any (and not only commercial) production or reproduction of the variety is subjected to the rights-holder's authorization (Option 1), a private-use exemption, as also required by that Convention, should be considered. If, as is the case under UPOV 1978, only commercial activities are covered by the *sui generis* right, there is no need for a private-use exemption.

Paragraph 2 (Research): As UPOV 1991 provides for a very broad, almost patent-like, scope of protection, it also includes — as does patent law — a research exemption (Option 1). In patent law, the interpretation of the research exemption varies from country to country. For example, in the United States it appears that the owner of a patented variety, or of a gene that has been inserted into a variety, can prohibit any use of the variety irrespective of the purpose. This prohibition prevents a patented variety from being used to create a new plant variety, even if the new variety does not include any protected material. In Europe, on the other hand, the patent-law research exemption is wider. For example, a researcher would probably not be prohibited from using patented plant material in research and development aimed at producing improved varieties, and commercial exploitation of the resulting varieties could not be prevented solely because they had been originally derived from patented material.

Paragraph 3 (Farming): This section addresses the so-called farmers' exemption, or the right to save seed. The farmers' exemption, in any or all of its possible forms, is one of the most commonly cited elements of Farmers' Rights, and yet there is no agreed-upon definition of Farmers' Rights.[26] The farmers' exemption could be limited to specific classes of farmers, such as small-scale farmers. Limiting the class of farmers who would qualify for the exemption is one possible way to strike a compromise among disparate interests: those of industry, concerned not to lose a significant portion of its market through the farmers' exemption; those of small-scale farmers, who want to foster the continued used of more heterogeneous varieties that continually adapt to specific local conditions; and those of agriculture-based communities, who want to continue culturally significant practices without interruption. (See the commentary to Article 10 on the perceived merits and demerits of heterogeneity and uniformity.)

Limiting the farmers' exemption to small-scale farmers is an approach taken in the Council Regulation (EC) No. 2100/94 of 27 July 1994 on community plant variety rights, and also in Brazil.

Paragraph 3, Option 1: The farmers' exemption may be considered as an element of Farmers' Rights in the International Undertaking on Plant Genetic Resources for Food and Agriculture (IU on PGRFA). Option 1 reflects the farmers' exemption as allowable under UPOV 1991. This definition of the farmers' exemption clearly rules out the so-called exchange of seed over the fence.

It might also prohibit the practice of explicitly allowing for the sale of propagating material harvested from protected varieties grown on a farmer's holdings (sometimes called 'brown-bagging').

Paragraph 3, Option 2 is consistent with Article 5.1 of UPOV 1978, under which authorization is required only for production for purposes of commercial marketing, offering for sale, and marketing of the reproductive or vegetative propagating material. By focusing on the commercial marketing of propagating material, UPOV 1978 implicitly allows the production of propagating material of a protected variety for non-commercial purposes. However, the scope of this so-called farmers' exemption or farmers' privilege is far from clear. While most UPOV member states have interpreted this as allowing farmers to plant-back seeds and to exchange limited amounts 'over the fence' on a strictly non-commercial basis, other member states, especially the United States, have interpreted the farmers' exemption as allowing farmers not only to replant seeds but also to sell limited quantities of them for reproductive purposes (brown-bagging). As recently as January 1995, the US Supreme Court held, in *Asgrow Seed Co. v. Winterboer*,[27] that under the farmers' privilege foreseen in the US Plant Variety Protection Act (PVPA), a farmer may sell for reproductive purposes only that amount of seed he or she has saved for replanting of his or her own acreage.[28] In other words, since that time, US farmers have not been allowed to produce or save seed specifically for reselling unless the amount is not more than needed for replanting their own acreage.

The farmers' exemption as defined here seems clearly to contradict UPOV 1991, unless the state allows exchanges 'over the fence' for reasons of 'public interest' (Article 17.1) and takes all measures necessary to ensure that the breeder receives 'equitable remuneration' (Article 17.2). Note that current US legislation, unlike that of Europe, does not include provisions to ensure that breeders receive equitable remuneration from farm-saved seed.

Paragraph 3, Option 3 explicitly allows for the sale of propagating material harvested from protected varieties grown on a farmer's holdings, or brown-bagging. It should be noted that while allowing brown-bagging in principle, some limits could be applied so as to allow the selling of certain quantities only, for example, a quantity equivalent to the farmers' own plantings, or a quantity less than that which farmers sell for other non-propagating purposes.

Viewpoint box 14: Who are small-scale farmers, and from what obligations should they be exempt?

Small-scale farmers need freedom.

The definition of 'small-scale farmer' is vital to the debate over 'farmer plant-back' and is possibly the most emotive issue in agricultural intellectual property discourse. Should it depend on income, on the area of land worked, or on the crop grown? There is growing acceptance that 'small-scale farmers' should be unrestricted in their ability to save and exchange seed — including intellectual property–protected seed — so long as a very limited proportion of this harvested seed enters the 'market' for replanting by others. Some are willing to accept farmer plant-back so long as a clear and limited definition of 'market' can be established. Others argue that the solution lies solely in the definition of 'small-scale farmer' and the proportion of the harvested commodity set aside for personal use. Legislators should at least agree that small-scale farmers have the right to save, improve and exchange (including sell) any seeds in their possession within their customary market area.

Markets need intellectual property protections.

Intellectual property law should not (and usually does not) interfere with private concerns. Subsistence farmers should be outside its scope, but not those for whom farming is a business. If the term 'small-scale farmers' is defined too widely, society will suffer because the incentive to improve seed will decline. 'Small-scale farmers' may be privileged to save proprietary seed for replanting and they may be permitted a local research exemption to adapt the seed to their immediate growing needs. They should never be allowed to sell proprietary seed.

Giving farmers the right to sell seed of a protected variety is controversial. Proponents of the farmers' exemption argue that traditionally farmers have always sold seed from their own harvest. That is undoubtedly true, but traditionally those varieties were not protected. Opponents argue that if farmers want to continue selling seed, they are free to cultivate non-protected (e.g., local) varieties. Commercial plant breeders, and more recently plant breeders in the public sector of several countries, have strongly opposed any exemption that allows any seed of protected varieties to be sold. They believe that such an exemption would not only jeopardize their rights, but render them null and void.

Paragraph 4 (Breeding) addresses the so-called breeders' exemption. Under UPOV 1978, authorization by the breeder is not required either for the use of the variety as an initial source of variation for the purpose of creating other varieties, or for the marketing of such other varieties (Option 1). The 1991 UPOV Convention limits this exemption insofar as the marketing of newly bred varieties requires the authorization of the owner of the original plant if these newly bred varieties have been essentially derived from the original plant. (See the definition of 'essentially derived varieties' in Article 5.)

Paragraph 5 (Local origin) provides for an exemption in cases where the protected variety in question has been derived from germplasm collected either in the territory of the state or the territory of those who shall benefit from this exemption. Some Crucible Group members believe that this option establishes a compulsory licence on a discriminatory basis and is therefore incompatible with the national treatment principle under TRIPs. Others argue that since the exemption only differentiates between residents and non-residents, it would still be in line with TRIPs' principle of national treatment. There is agreement that although the provision is worth further consideration, it may disadvantage local breeders, since in many cases their plant varieties might be derived from or based on germplasm collected in the country. There are some who think the local-origin exemption cannot be implemented because the 'percentage' of local material in a newly developed variety is not easy to define.

Paragraph 6 (Customary use): The customary-use exemption is still more contentious than the one on local origin. Option 1 might be unworkable or simply arbitrary if 'customary use' is not defined. If a clear and narrow definition of customary use is not given, this exemption could nullify any Plant Breeders' Rights, as some Crucible Group members fear. They believe it is important to add that the objective of this exemption should not be to broaden the scope of the farmers' exemption as defined in Paragraph 3. Others would like to see the whole paragraph dropped.

Article 18: Duration of rights

Option 1

The plant variety protection shall last for 20 years from date of grant of rights; 25 years for trees and vines.

Option 2

The plant variety protection shall last for 15 years from date of grant of rights; 18 years for trees and vines.

Option 3

The plant variety protection shall last:

a) in the case of plant varieties that are distinct, uniform and stable, for [number] years;
b) in the case of plant varieties that are novel, useful and non-obvious, for [number] years; and
c) in the case of plant varieties that are distinct and identifiable, for [number] years.

Commentary

The duration of rights can vary and is dependent upon a number of factors. It is customary, for example, to provide longer periods of protection for trees than for flowers, because trees take years to reach fruiting maturity. Creating a *sui generis* system is a juggling act requiring legislators to weigh and balance the scope of protection, rights granted, the duration of those rights, and the range and conditions of exemptions from those rights.

One standard administrative-law principle is that stronger, more exclusive rights should have a shorter duration than weaker, less exclusive rights. For example, the right to completely prevent third parties from reproducing protected varieties should expire before a weaker right such as the right to attribution. The right to attribution could last forever without seriously impairing the free flow of knowledge.

This way of thinking about the duration of the rights is very similar to the principle expressed above (in the commentary to Article 15), that the strength of the rights conferred on knowledge-holders should be inversely proportional to the breadth of the plant material that one is protecting. Of course, this formula — the more restrictive the rights, the shorter their duration — is based on the idea that the free flow of breeding material, on equitable and fair conditions, must remain the ultimate goal. Policy-makers may want to take competing goals into consideration.

The first two options are taken directly from the UPOV agreements. The third option provides different rights and durations for different groups of plant varieties. TRIPs does not set any minimum standards as to the duration of protection for plant varieties.

Option 1 is compliant with TRIPs and Article 19 of UPOV 1991.

Option 2 is compliant with TRIPs and Article 8 of UPOV 1978.

Option 3 addresses the possibility that a government might want to institute two or more parallel systems of protection.

Again, readers are reminded that the term 'useful' (in (b) above) requires some careful consideration. See Article 18 in Topic 3, Section 3 below for a discussion of why such a criterion could be difficult to implement in the context of an intellectual property protection law.

Article 19: Exhaustion of the rights

Option 0
no provision

Option 1
The rights conferred in Article 15 shall not extend to acts concerning any material of the protected variety [or a variety covered by Article 16, Element 5] that has been sold or otherwise marketed by the holder or with his or her consent in the territory of this country, or any material derived from the said material, unless such acts:

a) involve further propagation of the variety in question; or
b) involve an export of material of the variety, for further propagation, into a country that does not protect varieties of the plant genus or species to which the variety belongs.

Commentary

This provision addresses exhaustion of rights. It is an important principle of intellectual property law that you cannot have your cake and eat it too. Thus, a rights-holder given the sole right to commercial exploitation of a product may exercise the right by selling the product. The seller's rights in that product are then said to be 'exhausted'. However, the buyer is not completely free to act: what he or she can do depends on the exact nature of the rights. Under copyright laws, the buyer of an article or print-out does not have the right to make copies; the purchaser of a copy of a play cannot perform it in public. The purchaser of a patented article can use it commercially, repair it (but not replace it) and sell it; he or she may or may not be able to export it (depending on local law and where the patentee has other patents). Plant variety laws generally make sure that rights-holders can only exercise their rights and receive remuneration once in every production cycle. Exhaustion provisions are particularly important when the rights go beyond those foreseen in UPOV 1978 (commercial production/reproduction).

Option 1 draws on Article 13 of UPOV 1991.

Article 20: Compulsory licences

Paragraph 1: Conditions for compulsory licence

Option 0

no provision

Option 1

Any person may, by means of an application filed with the competent authority, seek the grant of a compulsory licence in respect of the rights conferred under this Law. A compulsory licence shall not be granted unless all the following conditions are fulfilled:

Element 1

the compulsory licence is necessary to safeguard the public interest

Element 2

the compulsory licence is necessary due to insufficient supply by the rights-holder of the subject variety

Element 3
the compulsory licence is necessary because too much of the plant variety being offered for sale by the rights-holder in this country is being imported rather than being produced in this country

Element 4
the rights-holder has refused to grant the applicant a licence on reasonable commercial terms

Element 5
the rights-holder is not prepared to grant a licence on reasonable terms

Element 6
three [or other number] years have elapsed between the date of the grant of the rights and the date of the application for the grant of a compulsory licence

Element 7
the applicant for the grant of a compulsory licence has paid the fee prescribed in the regulations for such grant

Element 8
the process of granting a compulsory licence and any licence so granted complies in all respects with Article 31 of TRIPs

Paragraph 2: The competent authority

Option 0
no provision

Option 1
The competent authority shall act in compliance with the general laws and regulations in granting or refusing a licence. In particular, it shall:

Element 1
offer the rights-holder the opportunity to be heard and to respond to all assertions of the applicant for the licence

Element 2
before granting or refusing a compulsory licence, hear interest groups and national professional organizations active in the field concerned

Paragraph 3: Appeals

Option 0
no provision

Option 1

Any decision of the competent authority to grant or refuse a licence, or on the amount of equitable remuneration therefor, shall be subject to appeal to a higher judicial authority. Both the applicant and the rights-holder shall have the right of appeal.

Commentary

With the exception of Paragraph 1, Element 8, the options set out in this article are relatively straightforward and do not require extended commentary or explanation. Paragraph 1, Element 8 states that compulsory licence provisions in the *sui generis* plant variety protection provisions anticipated in this text should comply with the TRIPs restrictions on compulsory licences. Commentators disagree over whether or not this is true. Some say that the application of Article 31 to Article 27.3(b)–compliant legislation (regarding *sui generis* intellectual property protection for plant varieties) is clearly intended in the spirit of TRIPs and, therefore, countries would be bound to limit their compulsory licence provisions to comply with Article 31. Others contend that a *sui generis* law to protect plant varieties could have much stronger compulsory licence provisions than those anticipated by Article 31 and still satisfy the 'effectiveness' criterion set out in Article 27.3(b), and that the spirit of TRIPs would be satisfied as a result.

Article 21: Anticompetitive practices

Option 0

no provision

Option 1

The following shall be prohibited: all agreements between undertakings (e.g., businesses, enterprises, traders, etc.), decisions by associations of undertakings and concerted practices which may affect trade and which have as their object or effect the prevention, restriction or distortion of competition within the market, and in particular those which:

a) directly or indirectly fix purchase or sale prices or any other trading conditions;
b) limit or control production, markets, technical development or investment;
c) share markets or sources of supply;
d) apply dissimilar conditions to equivalent transactions with other trading parties, thereby placing them at a competitive disadvantage; or
e) make the conclusion of contracts subject to acceptance by the other parties of supplementary obligations that by their nature or according to commercial usage have no connection with the subject of such contracts.

Any agreements or decisions prohibited pursuant to this Act shall be automatically void.

Commentary

Without doubt, this article goes far beyond the context of a plant variety protection law. The matter dealt with here is usually addressed by competition law. It has nonetheless been included here in order to draw readers' attention to the need for laws preventing anticompetitive practices. There is general agreement that any intellectual property legislation requires a law preventing abuse of the rights granted. The following agreements and practices are generally considered objectionable under this head:

a) an obligation on the licensee not to use the licensed variety or any related expertise after termination of the agreement, even though the expertise is not secret and the rights conferred under this law are no longer in force;
b) an obligation on the licensee to continue paying the royalties over a period going beyond the duration of the rights conferred under this law; and
c) an obligation on the licensee not to supply, or to supply only a limited quantity of, material of the variety to a particular customer.

Other agreements and practices that may be considered objectionable in some circumstances include:

d) an obligation on the licenser not to license other undertakings to exploit the licensed plant variety in the licensed territory;
e) an obligation on the licenser not to exploit the licensed plant variety in the licensed territory him- or herself;
f) an obligation on the licensee not to market material of the protected variety in territories of countries that are licensed to other licensees;
g) an obligation on the licensee not to grant sublicences or to assign the licence;
h) an obligation on the licensee to assign or license to the licenser her/his own improvements of the variety;
i) an obligation to take or to assist the licenser to take legal action against misappropriation or infringements of the rights conferred under these provisions; and
j) an obligation on the licenser to grant the licensee more favourable terms than the licenser may grant to another undertaking after the agreement is entered into force.

While all Crucible Group members agree on the need for anti-trust legislation (see Recommendation 4, 'Anti-trust legislation for seed industry', in *Seeding Solutions,* Volume 1, p 17), several feel strongly that anticompetitive practices should be dealt with in separate legislation. Further, there is only limited agreement on which practices should be forbidden. In particular, some group members consider that practices (d) to (j) are often, perhaps even usually, acceptable. Thus, they argue, (d) and (e) are normal terms in any exclusive licence, and are only objectionable if combined with (g), since the result of the three terms combined is that further licensing is impossible. Similarly, (g) by itself is a normal and completely acceptable provision in a non-exclusive licence, while (f) is no more than a territorial restriction on licensing, which should be quite acceptable except in special circumstances. Practices (i) and (j) are frequently included in licensing terms, and are not necessarily oppressive depending on the relationship of the parties. Whether (h) is acceptable also depends on the particular situation, including the status of the parties and the value to each of the technology licensed. Unnecessary restrictions on licensing terms, it is argued, are to be avoided: they will reduce the incentive to license and slow the transfer of useful technology, making everyone worse off.

Article 22: Derogation from exemptions

Option 0
no provision

Option 1
Private contracts shall be unenforceable to the extent that they impose restrictions on the use of material of the variety or parts thereof, in conflict with the exemptions set out in Article 17.

Commentary
The purpose of this article is to prevent breeders and farmers from circumventing important policy decisions taken by this law by concluding private contracts that exclude activities otherwise allowable under this law. While some members of the Crucible Group believe that this article is essential, since otherwise the exemptions provided by Article 17 are so easily evaded as to have little or no effect, other members feel that the law should hesitate to interfere with private contracts except in extraordinary circumstances.

Article 23: Persons entitled to file applications

An application for a right conferred under this Law may be filed by:

Option 1
any persons (natural or legal).

Option 2
any persons (natural or legal) resident, or foreign nationals having a registered office in [the implementing country].

Option 3
any persons (natural or legal) resident, or foreign nationals having a registered office in [the implementing country] if, with respect to Plant Breeders' Rights laws in the home countries of the foreign nationals, the nationals of [the implementing country] enjoy the same treatment as those foreign nationals.

Commentary

This article addresses the question who may apply for plant variety protection within the implementing country. All options give the right to apply to natural and legal persons (i.e., corporations).

Only Option 1 allows all natural and legal persons to apply for the rights. This is the most liberal option. Option 2 is slightly more restrictive, limiting application to those who are either resident or have their registered office in the country. Options 1 and 2 conform to TRIPs since they do not discriminate based on the applicant's nationality. Option 3 reflects UPOV 1978 by applying the principle of reciprocity. The principle of reciprocity means that foreigners and foreign companies enjoy the same treatment as the nationals of the country only if their home countries' plant variety protection laws also acknowledge this principle of reciprocity. It should be noted that this option is not TRIPs-compatible, because the principle of national treatment, which also applies to the *sui generis* system for the protection of plant varieties, does not allow for reciprocity. Thus, WTO members implementing the minimum requirements set by UPOV 1978 would not satisfy the TRIPs requirements with regard to plant variety protection.

Article 24: Examination of applications

Option 1

The variety shall undergo, under the supervision of the competent authority, an examination based on results obtained by the applicant.

Option 2

The variety shall undergo, under the supervision of the competent authority, a technical examination including growing tests or other necessary tests to be carried out by the competent authority.

Commentary

In **Option 1**, the competent authority reviews data (e.g., on distinctness, uniformity, or other appropriate criteria) supplied by the applicant in order to decide whether protection is to be granted.

Under **Option 2**, the authority itself grows samples of the seed supplied, makes its own observations and decides based on these.

A combination of Options 1 and 2 might be the best solution in most cases.

Article 25: Denomination

The denomination of a protected variety shall be its generic designation. The denomination may consist of any suitable word, combination of words, combination of words and figures or combination of letters and figures, provided that the denomination allows the variety to be identified.

Commentary
Some Crucible Group members wish to point out that within an international protection system such as UPOV, varieties should always be commercialized under one and the same name in all of the systems' member states.

Article 26: Competent authority

Option 1
The national patent office shall be the competent authority.

Option 2
The national plant variety protection office shall be the competent authority.

Option 3
The national authority for the certification of seeds shall be the competent authority.

Option 4
Additional administrative bodies [e.g., a farmers' trust, a plant breeders' or small-scale farmers' ombudsman] shall be set up if required.

Commentary
This article lists new administrative offices that may need to be set up. However, no details regarding these offices are provided.

Option 1: The implementing country might choose the already existing patent office as the competent authority.

Option 2: Alternatively, the country may wish to establish a new plant variety protection office, perhaps with links to the ministry of agriculture.

Option 3: A third alternative open for countries which already have or intend to establish a seed certification system is to have one authority which is responsible both to certify seeds and to grant rights under this law. In fact, this is the current practice in some UPOV member countries such as Germany.

Option 4: Finally, additional administrative bodies could be established, but their function should be defined as clearly as possible.

Part Six: Interface

Article 27: Interface with seed certification system

Option 0
no provision

Option 1
The rights conferred under this Law shall be independent of any measure taken by the state to regulate the production, certification and marketing of material of varieties or the importing or exporting of such material.

Article 28: Interface with patent law

Option 0
no provision

Option 1
Notwithstanding any patent rights restricting the use of material of the variety or parts thereof, the exemptions as set out in Article 17 shall apply.

Commentary
This article addresses the relationship between the rights granted under this provision and patent rights. This provision may only make sense for those countries that do not opt for the patentability of plant varieties. For such countries, the situation may arise that a plant protected under this law could include certain genetic information subject to a patent. The question then is whether the use of material of this plant variety is governed by this provision, by the patent granted for the genetic information included in the plant, or by both. This article aims to ensure that the exemptions chosen under Article 17 do not lose their relevance where the protected plant variety includes patented material. However, some Crucible members maintain that the grant of plant variety rights should not be used to abridge the exclusive rights of patentees, if those rights are otherwise justified.

See 'The relationship between patent and other rights (e.g., plant variety protection)' in the Appendix for a more detailed discussion of this issue.

Section 3

Options for intellectual property laws for biotechnological innovations

Contents

Section 3

Options for intellectual property laws for biotechnological innovations

Introduction

Crucible members are divided about the value of intellectual property laws. Some regard them as essential, or at least desirable, for all countries, to promote innovation and development and reward innovators fairly. Others doubt their value in almost all situations (literary copyright apart), considering this at best unproven and at worst pernicious. Others admit their value in some circumstances, but deny them in others: for example, they find them ill adapted to promoting biological innovation, or to the special needs of developing countries.

However, all Crucible members recognize the significance of the Trade-Related Aspects of Intellectual Property Rights (TRIPs) Agreement for developing countries. TRIPs requires members to provide intellectual property protection, generally in the form of patents, for most forms of innovation, including microbiological products and processes, and plant varieties. Many countries will wish to join the World Trade Organization (WTO), because of the trading advantages it offers them. Therefore, such countries, whatever view they take about the intrinsic value of intellectual property, will wish to introduce intellectual property laws conforming with TRIPs. Countries who decide they do not wish to join the WTO may nevertheless decide to introduce some form of intellectual property protection for biological inventions. It will be important to know, however, whether any proposed provisions conform to TRIPs or not. In what follows, some options may be incompatible with the obligations that WTO members have under TRIPs. These options have been identified throughout the text. Members (or aspiring members) of the WTO should carefully consider the implications of selecting such options.

TRIPs requires the grant of patent rights for all inventions meeting certain conditions, including novelty, inventiveness and practical utility. TRIPs does not prohibit the grant of additional rights in other inventions that do not meet these conditions. TRIPs sets minimum standards, but it does not prevent countries from granting additional intellectual property rights. Further, TRIPs creates minimum obligations for member states to protect all kinds of inventions, including certain biological inventions relating to, for example, microorganisms and microbiological processes, but also allows other biological innovations, such as plants and animals, to be excluded from patentability.

Why do we give such emphasis to TRIPs? Is it set in stone? Crucible members differ widely on this. Some of us believe that TRIPs is fundamentally unjust and must be amended, no matter how difficult that may be to do. Others see TRIPs in its present form as the result of careful and hard-fought negotiation. They see no clear injustices, and even if the result is less than perfect, the political situation is such that in the next several years it will be practically impossible to obtain consensus for significant changes. We agree only that, in theory at least, it is possible to amend TRIPs, and that any such amendment will be exceptionally challenging to bring about. We counsel readers to bear both these points in mind.

The important questions for a law of this kind include:

What kinds of subject matter can be protected?

(processes? products? microbes? plants? animals?)

What are the rights of the inventor?

(to exclude others? a royalty? an acknowledgement of authorship?)

What exemptions should be provided to such rights?

(research use? private use? use by farmers/indigenous people? none?)

Should discoveries be protected, or only inventions?

(and what is the difference?)

Should inventions using biological materials require special conditions?

(Certificate of origin? Prior informed consent?)

These choices and others are treated in the text that follows.

Part One: Purpose and scope

Article 1: Purpose

The purpose of this Act is:

Element 1

to determine what kinds of biological innovations may be protected by patents and under what conditions

Element 2

to encourage the generation, development and exploitation of biological innovations for the benefit of the public

Element 3

to regulate the ownership of biological innovations to prevent exercises against the public interest

Element 4
to share benefits equitably between the suppliers of biological resources and those who use them as the basis of innovations

Element 5
to exclude from patent protection innovations whose exploitation must be prohibited as contrary to *ordre public* or morality, or as liable to seriously damage the environment

Element 6
to exclude from patentability biological material and processes

Element 7
to promote innovation in biotechnology while ensuring the conservation and sustainable use of biological diversity and respecting, preserving and maintaining knowledge, innovations and practices of indigenous and local communities embodying traditional lifestyles relevant for the conservation and sustainable use of biological diversity[29]

Commentary
This list provides the most likely purposes or objectives of intellectual property legislation regarding biological innovations. Certainly not all possibilities are included here. It is not necessary to have a purpose section. Together with the preamble, however, a purpose section can be a valuable interpretive device to assist in discerning the meaning of ambiguous sections throughout the body of the legislation.

Element 1 is neutral and does not set any clear objective.

Element 2 emphasizes the positive benefits of protection.

Element 3 emphasizes the need to protect the public interest.

Element 4 introduces the idea of equitable benefit-sharing.

Element 5 mentions specifically exclusions allowed by TRIPs.

Element 6 rejects the idea of any 'patents on life'.

Element 7 combines Elements 2 and 5.

All these elements are likely to be TRIPs-compliant, except Element 6, which clearly breaches it. Element 7 also promotes the objectives of the Convention on Biological Diversity (CBD).

Article 2: Scope

This Law shall apply to all inventions that relate to, make use of, or consist of biological material.

Commentary

This provision defines the scope of the law. The term 'invention' requires some explanation, which is given in Article 3 below.

Article 3: Invention/discovery

Option 1

An invention shall offer a technical teaching for a technical problem.

The invention must be disclosed in writing in a manner sufficiently clear and complete for it to be carried out by a person skilled in the art.

Naturally occurring substances are discoveries and, therefore, excluded from patentability.

Option 2

An invention shall offer a teaching to methodically use controllable natural forces to achieve a causal, perceptible and repeatable result.

The invention must be disclosed in a manner sufficiently clear and complete to be carried out by a person skilled in the art. If an invention concerns or involves the use of biological material which is not available to the public and which cannot be described in the patent application in such a manner as to enable the invention to be carried out by a person skilled in the art, the invention shall be regarded as being disclosed if a sample of the biological material has been deposited with a recognized depository institution not later than the date of filing of the application.

Discoveries are not patentable. If, however, a substance found in nature has first to be isolated from its surroundings, and can be properly described either by its structure, by the process by which it is obtained, or by other parameters, that substance is patentable.

Option 3

An invention is any new and useful process, machine, manufacture or composition of matter, or any useful improvement thereof.

Commentary

To indicate what can and cannot be patented, some national patent laws include definitions of 'inventions'.

It is extremely difficult to define the term 'invention'. In fact, most patent laws do not even try to define it. Neither TRIPs nor the Paris Convention for the Protection of Industrial Property give any definition of what an invention should be. US patent law (see Option 3) only gives a definition of what may be invented, i.e., any new and useful process, machine, manufacture or composition of matter, or any useful improvement thereof. The term 'invention' is nonetheless crucial for the debate on the patenting of biological innovations. This debate focuses on three features of 'inventions': (1) their technicality, (2) their reproducibility, and (3) the difference between 'invention' and 'discovery'.

The three options offered here reflect two schools of thought. Option 1 reflects the thinking of those who basically deny the applicability of patent law to biological innovations. Options 2 and 3 reflect the thinking of those who are in favour of patenting biological inventions. Option 3, derived from the US patent statute, represents a 'common law' way of thinking about inventions: that they are new physical objects, materials or processes, rather than 'teachings'.

There are other requirements that inventions must meet to be patentable: for example, they must be 'new'. We will examine those conditions in Part Two, 'Conditions for the grant of protection', below.

Options 1 and 2 address, in their first parts, the question whether biological material may be the subject of an invention. While Option 1 defines invention in a technical manner that might exclude biological innovations from patentability, Option 2 regards biology as a 'natural force' equivalent to chemistry or physics.

In their second parts, these options address the issue of repeatability of an invention, which is closely related to the invention's disclosure. Both options assume that inventions need to be repeatable. However, while Option 1 assumes that an invention is repeatable only if it can be sufficiently described in writing, Option 2 creates a special rule for inventions involving biological material. It may not always be possible to describe in words innovations that incorporate biological material. The solution, therefore, is to create a depository for samples of biological material and state that this deposit satisfies the disclosure criterion. A person seeking disclosure about a deposited invention could simply obtain a sample of the deposited material.

Options 1 and 2 both address the issue of discoveries. Again, they acknowledge that discoveries should not be recognized as inventions. However, Option 1 defines all naturally occurring substances as discoveries. Genes and gene sequences isolated from a naturally occurring organism could not, therefore, be patented under this option. In contrast, Option 2 excludes only those freely occurring substances that someone merely found. If, however, the substance had first to be isolated from its surroundings, and can be properly described, it shall be patentable whether that substance existed in nature before it was recognized or not.

Option 3 adopts a different approach to discoveries. There is no ban on patenting discoveries as such. However, to be patentable, a discovery must

Viewpoint box 15: Should discovery be distinguished from invention in patent laws?

Discovery differs from invention.

The distinction between 'discovery' and 'invention' is crucial to patent law. The European Patent Convention (EPC) explicitly excludes 'discoveries'. In the United States, in the 1948 case of *Funk Bros. Seed Co. v. Kale Inoculant Co.*, a patent for a mixture of nitrogen-fixing *Rhizobium* bacteria was denied on the grounds that it was a discovery of a phenomenon of nature. The US Supreme Court held that "the qualities of these bacteria, like the heat of the sun, electricity, or the qualities of metals, are part of the storehouse of knowledge of all men. They are manifestations of laws of nature, free to all men and reserved exclusively to none." This so-called product-of-nature doctrine was reaffirmed in the landmark 1980 case of *Diamond v. Chakrabarty*, in which the Supreme Court emphasized that the genetically modified oil-consuming bacteria in that case were a 'manufacture' rather than an unpatentable product of nature. However, due to constant pressure from concerned interest groups, the line separating invention from discovery has become very thin under both US law and the EPC. Naturally occurring substances, like cells or genes, have become patentable under US law if they have been 'purified or isolated'. The same is true in the EU despite the EPC's exclusion of discoveries.

Despite these trends in the law, patents should be awarded only for specific expressions of human ingenuity, not for the revelation of something that already exists. Allowing patents on materials — genes, for example — that are simply discovered or isolated from their natural surroundings — often using routine methods — unjustly monopolizes material that already is, and should remain, in the public domain. Some argue that it does not really matter whether you discover or invent something as long as it is beneficial. This argument, however, reveals a fundamental misunderstanding: it is not, nor has it ever been, patent law's function to grant protection for whatever is beneficial. Einstein's theories have never been patented and yet may be the most important scientific breakthroughs the last century has seen. Furthermore, methods of therapeutic treatment of human beings are of utmost importance and truly beneficial, but they are still excluded from patentability in Europe.

Discovery leads to invention.

US law says (35 USC 100): "'Invention' means 'invention or discovery'". Although European law says the opposite, what you can protect is much the same under either law. This proves that the distinction between 'discovery' and 'invention' is not fundamental.

The word 'discovery' is being used in slightly different ways in the two laws. For Europeans, a 'discovery' (English law talks of 'a mere discovery') is no more than unpatentable information: e.g., the discovery of a new gene sequence that causes, say, resistance to mildew in a particular known variety of wheat. If this 'discovery' could be patented, the known variety of wheat could be taken out of the public domain. That would be fundamentally wrong, hence the prohibition. However, for US lawyers, no such prohibition is necessary — if any patent covers something already known, it is invalid for lack of novelty. Therefore, they do not hesitate to use the word 'discovery' for such a gene sequence, and do not regard claims based on it as unpatentable (e.g., to an isolated gene sequence, a construct or plant containing the recombinant sequence, etc.), provided these do not cover anything old or obvious. In practice, European law has the same effect and allows the same claims. However, the European prohibition on patenting 'discoveries' leads some to say (wrongly) that a patentable invention (a new product or process) cannot be based on an unpatentable discovery (new information about something already existing). In fact, this is a typical route to invention, particularly in the chemical and biological areas.

From the point of view of the public interest in bringing beneficial new things into use, the distinction between discovery and invention is irrelevant. Suppose there is a new substance that protects from malaria without side effects: what difference does it make whether the substance has been newly synthesized in a laboratory or discovered pre-existing in a marine organism?

be material (rather than, say, a law of nature) and it must also be 'new'. This leaves open the question of when something pre-existing in nature can be considered 'new'.

If Option 1 is understood to exclude all living material from patentability, it is clearly incompatible with TRIPs. The TRIPs Agreement explicitly prohibits the exclusion of microorganisms from patentability: the mere fact that an invention consists of, or relates to, biological material cannot render the invention unpatentable (see Article 5, below). With regard to depositing biological material and to excluding naturally occurring substances, including genes and gene sequences, Option 1 might be in accordance with TRIPs. The TRIPs Agreement does not oblige WTO members to adopt a deposit system for biological inventions. Nor does TRIPs include any reference to the 1980 Budapest Treaty on the International Recognition of Deposit of Microorganisms for the Purposes of Patent Procedure.

Option 2 is consistent with TRIPs. A different question is whether the patenting of naturally occurring biological material violates any obligations under the CBD. Some would argue that to patent naturally occurring genetic resources collected after the CBD entered into force and without the consent of the country of origin (assuming that the country has a national access law requiring consent) would violate the Convention.

Option 3 is in line with TRIPs, provided 'composition of matter' is interpreted broadly to include organisms.

Article 4: Exemptions relating to *ordre public* or morality

Option 0

no provision [i.e., there is no exemption]

Option 1

Patents shall not be granted for inventions when publication or exploitation of them would be contrary to *ordre public* or morality, provided that such exemption is not made merely because the publication or exploitation is prohibited by other laws.

Option 2

Patents shall not be granted for inventions when prevention of commercial exploitation of them within this country's territory is necessary to protect *ordre public* or morality, including to protect animal or plant life or health or to avoid serious prejudice to the environment, provided that such exemption is not made merely because the exploitation is prohibited by other laws.

Commentary

Option 1, reflecting EPC Article 53(a), exempts inventions which are contrary to *ordre public* or morality.

Option 2 restricts Option 1 to inventions whose exploitation within the territory needs to be banned for such reasons.

Some members of the Crucible Group are of the opinion that Option 1 breaches TRIPs (unless its application is limited to subject matter, such as animals, that TRIPs allows to be excluded). Others believe that EPC Article 53(a) is in line with TRIPs Article 27.2.

Option 2 allows inventions to be refused patent protection only where the problem is so serious that it is also necessary to ban their exploitation. This specifically conforms to TRIPs Article 27.2. This means that subject matter cannot be declared unpatentable simply because the act of patenting as such is considered offensive. Only if commercial exploitation must be prevented may an exemption based on morality or *ordre public* be made. Industry generally prefers Option 2 to Option 1, because it appears to be somewhat clearer.

The question of what offends against morality is contentious. Some contend that Options 1 and 2 are so restrictive that they will rarely be applied, thus allowing the patenting of, and giving direct encouragement to, many dubious activities. For the Crucible Group's debate on this subject, see the Viewpoint box entitled 'Can TRIPs' public morality exclusion be used to reject patents on life forms or controversial new technologies such as genetic seed sterilization?' in *Seeding Solutions*, Volume 1, p 90.

Article 5: Exemption of biological material

Option 0

no provision

Option 1

Patents shall not be granted for biological material.

Option 2

Patents shall not be granted for living organisms.

Commentary

This provision addresses the issue of patents on life forms. Whether or not patents on biological or living material should be granted is still controversial. Under Option 0, such material would not explicitly be excluded from patentability. Option 1 excludes all material derived from organisms (such as proteins and DNA) from being patented. Option 2, somewhat narrower, excludes living material from patentability. Options 1 and 2 are both clearly incompatible with TRIPs. The TRIPs Agreement, by explicitly prohibiting the exclusion of microorganisms from patentability, makes it very clear that the mere fact that an invention consists of or relates to living biological material does not render the invention unpatentable (see also the commentary to Article 3, 'Invention/Discovery').

Article 6: Exemption of human beings/parts of human beings

Option 0

no provision [i.e., there is no exemption]

Option 1

The human body, at the various stages of its formation and development, cannot constitute a patentable invention.

Option 2

An element isolated from the human body or otherwise produced by means of a technical process, including the sequence or partial sequence of a gene, may constitute a patentable invention, even if the structure of that element is identical with that of a natural element.

Option 3

Patents shall not be granted for the human body or parts thereof, including elements isolated from the human body, such as cells, subcellular structures, genetic material and biochemical substances.

Commentary

In this article, various options are offered with regard to the human body and parts thereof. The Crucible Group offers comments in this area with extreme diffidence. We are not experts in either medicine or ethics, two disciplines that have much to say about these problems. We are more concerned with the effects of legislation on genetic resources, especially for agriculture. However, the provisions of TRIPs oblige countries to consider these areas, and it will not do for us simply to pass them by as too difficult. Further, although this is a question to which experts must contribute, it is not one that they should be allowed to decide unaided.

Under **Option 0**, the human body and parts thereof would not be explicitly excluded from patentability. In fact, most patent laws do not exclude the human body or its parts from patentability. This, however, does not mean that the human body or its parts are necessarily patentable under these laws. It is questionable, for example, whether biological innovations relating to the human body would meet the patent requirement of industrial applicability, or, if they did, whether any resulting patents could be enforced.

Option 1 excludes patents on the human body. This seems to be compatible with TRIPs, although the agreement does not mention the human body as such. (It does allow animals to be excluded from patentability.)

Option 2: With regard to parts of the human body, Option 2, which reflects the European Patent Directive (98/44/EC), requires that in order to be patentable those parts have at least to be isolated from the human body. Once they have been isolated, they are patentable whether they are identical with naturally occurring elements or not.

Option 3 excludes patents on the human body as well as parts thereof. It is argued that such parts may be excluded from patentability pursuant to TRIPs, given that whole human beings can be excluded pursuant to TRIPs. But TRIPs does not explicitly consider human parts. Some might argue that TRIPs does not allow the exclusion of parts of human beings. For example, in 1990, the Supreme Court of California held in *John Moore v. the Regents of the University of California*[30] that an invention derived from cells of Mr Moore's body could be patented. Some argue that the *Moore* decision is right and that the possibility of excluding animals (including human beings) from patentability should not extend to human parts. They say that parts of the human body that are successively more remote from the human body as a whole — for example, genetic sequences and biochemical substances — are suitable patentable subject matter. Human parts vary in importance — brains are more important than toenails, which are more important than the water that makes up at least 60% of human bodies. A human being is very much more than the sum of its parts. Respect for human dignity requires respectful treatment of all human beings, but not necessarily equal respect for all materials that are derived from, were once part of, or are copied from, human beings. On the other hand, trade in human organs is regarded by many as immoral and as a violation of human dignity; in fact, it is punishable for this reason in several countries.

Article 7: Exemption of animals

Option 0

no provision [i.e. there is no exemption]

Option 1

Patents shall not be granted for animals.

Commentary

Option 1, excluding animals, is specifically allowed by TRIPs Article 27.3(b). It should be noted that there is no obligation under TRIPs to provide for *sui generis* protection of animals or animal races (*races animales* in the French version; *Tierarten* in the German version).

Article 8: Exemption of plants

Option 0

no provision [i.e., there is no exemption]

Option 1

Patents shall not be granted for plants other than plant varieties.

Option 2

Patents shall not be granted for plant varieties. Inventions that concern plants are patentable provided that the application of the invention is not technically confined to a single plant variety.

Option 3

Patents shall not be granted for plants.

Option 4

Patents shall not be granted for inventions relating *inter alia* to plant varieties belonging to species or genera for which protection is available under the law [for the Protection of New Varieties of Plants].

Option 5

Patents shall not be granted for innovations that can be protected under the *sui generis* law relating to the protection of plant varieties.

Commentary

This provision addresses plants and plant varieties. Under TRIPs, plants may be excluded from patentability. However, WTO members must provide for the protection of plant varieties "either by patents or by an effective *sui generis* system or by any combination thereof". This leads to quite a few options. In order to comply with the TRIPs requirement regarding plant varieties, WTO members may choose not to exclude plants from patentability (Option 0) or to exclude plants except plant varieties (Option 1). They may also choose to exclude plant varieties from patentability, but to grant patents for inventions that are not confined to a single plant variety (Option 2). They may choose to exclude all plants (including plant varieties) from patentability (Option 3). They could choose to provide patent protection only for those varieties that are not protectable otherwise, for example under the Plant Breeders' Rights legislation (Option 4). They could also choose to exclude from patentability whatever could be protected under their *sui generis* law for the protection of plant varieties (Option 5).

Option 0 allows for the patenting of plants. WTO members choosing this option would not have to establish a *sui generis* system for the protection of plant varieties.

Under **Option 1**, plants as such would be excluded from patentability. Plant varieties, however, would be offered patent protection. WTO members choosing this option would not have to establish a *sui generis* system for the protection of plant varieties.

Although Options 0 and 1 both clearly comply with TRIPs, they are nevertheless highly unpopular with commercial breeders, many of whom would prefer a system similar to that of the International Union for the Protection of New Varieties of Plants (UPOV). Viewpoint box 16, 'Should patents on plant varieties be allowed?', below, explores this argument in more detail.

Option 2 reflects the legal situation that will prevail in the member states of the European Union (EU) after Directive 98/44/EC is adopted. Under this Directive, plant varieties as such are excluded from patentability. However, if an invention is applicable to more than one single plant variety, the invention may be patented. In this case, the patent claims relating to 'plants' would, of course, also embrace plants belonging to plant varieties, although the latter are excluded from patentability.

Some people say this is illogical, even ridiculous. The European law contains a clear prohibition against patenting plant varieties (Article 53(b)). How can it make sense to allow patents on plants, while excluding patents on plant varieties made up of plants? Further, if one variety cannot be patented, how can two be? If there are to be patents in this area, how can it be just to discriminate against inventions that relate to a single variety only?

To others, there seems to be no problem. They accept the resolution of this issue in the *Re: Novartis*[31] decision of the EPO Enlarged Board (20 Dec 1999). In that decision, the EPO distinguishes between the subject matter of the rights and the scope of protection.

Patents are granted for 'inventions'. Certain categories of innovation are not patentable: for example, according to European law, 'ideas', 'computer programs', 'discoveries', and 'aesthetic creations' (EPC Article 52). But the fact that these innovations cannot form the basis of a patent grant is not to say that they cannot fall foul of someone else's patent. To say that a subject matter is not patentable is not necessarily to grant it an immunity against infringement suits (excluded from the scope of patent protection). There is a strong analogy between plant variety rights and registered rights for aesthetic

Viewpoint box 16: Should patents on plant varieties be allowed?

Paradoxically, providing only patent protection would result in less protection overall.

It is quite clear that the obligation to protect plants under TRIPs is fully met by providing the possibility of obtaining patent protection for plant varieties. Patents, however, need only be granted for inventions meeting the usual conditions: novelty, inventiveness, utility, repeatability. Until very recently, only a very few developing countries had any intellectual property system for the protection of plant varieties in place, and many of them are still doubtful about the wisdom of protecting plant varieties at all, and hence feel justified in offering minimum protection. Although it is often said that no domestic private industry would develop without intellectual property rights, the absence of such rights has often facilitated enormous economic wealth, which, after a period of 'copying', often leads to private research and development investments.

Paradoxically, minimum protection for plant varieties may be offered by providing only patent protection for them. This is because in many cases it may be difficult to meet reasonable criteria for the grant of patents: in particular, inventiveness and repeatability.

Plant varieties are typically produced by crossing and selection; the new varieties contain some traits of both parents. In principle, the process and result are often obvious, though they may take years of expensive and painstaking work. In such cases, a Patent Office may reasonably reject a patent application for lack of inventiveness. Furthermore, an invention, to be patentable, must be described in such a way as to enable others to repeat it. Breeding processes often are not repeatable from a written description: to reproduce the new variety, one may require access to biological material, e.g., parent lines or even the seed of the new variety itself. While some patent offices allow the deposit of biological material to assist patentees to describe their inventions in a reproducible way, there is no obligation under TRIPs to do this. A country that strictly applied these two requirements would probably grant very few patents for plant varieties.

This approach is not only perfectly consistent with TRIPs, it is an approach explicitly mentioned in TRIPs. If there is anything 'cynical' about this approach, then TRIPs is to blame for this cynicism, not the countries implementing it.

Perhaps, but this would be a cynical exploitation of a loophole.

If plant varieties were deemed patentable, there would be full compliance with TRIPs. However, many agree that the patent system is not well adapted to the protection of plant varieties. This was, after all, why the UPOV system came into being: to offer protection to breeders whose creations might not be protected under existing patent law, and to allow breeders and farmers some exemptions that had not been found necessary under existing patent law.

It is both short-sighted and cynical — a step back 30 years — to recommend offering patent protection only for plant varieties. In a patents-only system, a strict application of the criteria for patentability could deprive many important new varieties of all protection. This failure to provide effective protection might well be seen as a violation of the spirit of TRIPs. More importantly, it would send a message to commercial breeders that countries offering such limited protection were not interested in benefiting from their investment or expertise.

designs. Both relate to the appearance of the matter protected (in the case of plant variety rights, morphological character), and in each case, the rights have a limited scope, being avoided by relatively small changes. Aesthetic designs, like plant varieties, are subject matter excluded from patent protection; instead, as with plant varieties, a separate system of protection is provided for them (i.e., registered design laws). However, unlike with plant varieties, no one has ever suggested that having a registered design right would entitle someone to ignore someone else's patent.

As to injustice to the inventor of a single plant variety, this, it is argued, is unrealistic. A plant variety is not in itself an invention, although it may embody or exemplify an invention. If it does, it is extremely rare, or even unknown, for the invention to be uniquely relevant to a single variety. In exactly the same way, it is extremely rare for an invention to be uniquely applicable to a single registered design. Even if the claims are so limited, the underlying invention will be broader. Even supposing there could be a case in which an invention involves a single variety, the situation of an invention applicable to two (and only two) varieties simply does not arise. If there is a general inventive idea connecting the two varieties, this itself, and not just the two varieties, is the subject of the patent. If there is no such inventive idea, then the patent lacks unity: it is two separate inventions illegitimately posing as one.

Option 3 excludes plants, including plant varieties. It is perfectly in line with TRIPs. However, members choosing this option must provide for the protection of plant varieties through an effective *sui generis* system.

Option 4 distinguishes between plant varieties protectable under the country's Plant Breeders' Rights legislation and other plant varieties which are not. This option reflects the so-called 'ban on double protection' as stipulated by UPOV 1978.

Option 5 is similar to Option 2 — it represents one view of EU law to date.

Article 9: Exemption of parts of plants and animals

Option 0

no provision [i.e., there is no exemption]

Option 1

Patents shall not be granted for parts of [plants] [and] [or] [animals], such as elements isolated from them (for example, cells, subcellular structures, genetic material and biochemical substances).

Commentary

This provision addresses the patenting of parts of plants and animals. Countries have, of course, the option to exclude only parts of plants, or only parts of animals. For simplicity, and because they raise similar legal questions, both options have been included in Option 1.

It is not really clear whether the exclusion of parts of plants and animals is allowable under TRIPs. As with human parts, discussed in Article 6, it is said that if whole animals or plants are to be excluded from patentability, so should parts of them. Otherwise, a patent applicant could claim all parts of a plant instead of the plant as such. Again, this is contested by those who want to patent such things as genes, cells and proteins derived from animals and plants.

Article 10: Exemption of microorganisms

Option 0
no provision [i.e., there is no exemption]

Option 1
Patents shall not be granted for microorganisms.

Option 2
Patents shall not be granted for microorganisms, except transgenic microorganisms.

Commentary
This provision addresses the patenting of microorganisms and of parts thereof. Option 1 is clearly contrary to TRIPs, which does not allow microorganisms to be excluded from patentability. The same might apply to parts of microorganisms.

There is no obligation under TRIPs to recognize the deposit of microorganisms as a form of disclosure equivalent to a written description of the organism.

Option 1 excludes microorganisms from patentability and thus would not conform with TRIPs Article 27.3(b).

Option 2 would exclude from patentability only those microorganisms which have been found in nature and which have not been modified by the patentee. Since TRIPs is silent on the patentability of naturally occurring substances, some will argue that Option 2 complies with TRIPs.

Article 11: Exemption of pharmaceuticals

Option 0
no provision [i.e., there is no exemption]

Option 1
Patents shall not be granted for pharmaceuticals.

Commentary

This provision addresses one of the most controversial aspects of TRIPs, the obligation to provide patent protection for pharmaceuticals.

Option 1 is incompatible with TRIPs as currently drafted, except as noted in the next paragraph.

Where a WTO member did not made available, as of 1995, patent protection for pharmaceuticals and agricultural chemical products, this member state must establish a mechanism ready at any time thereafter for the grant of exclusive marketing rights (AB-1997-5, India: Patent protection for pharmaceutical and agricultural chemical products). Developing-country members that did not provide product patents on agricultural chemical products and pharmaceuticals before 1995 may continue to exclude them from patentability until 2005.

Article 12: Exemption of agricultural chemical products

Option 0

no provision [i.e., there is no exemption]

Option 1

Patents shall not be granted for agricultural chemical products.

Commentary

This provision addresses agricultural chemical products, which under TRIPs (as with pharmaceuticals) may be excluded, for a limited period only, by specific developing countries under specific conditions.

Article 13: Exemption of biological processes

Option 0

no provision [i.e., there is no exemption]

Option 1

Patents shall not be granted for biological processes.

Option 2

Patents shall not be granted for biological processes for the production of plants and animals.

Option 3

Patents shall not be granted for essentially biological processes for the production of plants and animals, other than non-biological and microbiological processes.

Option 4

Patents shall not be granted for processes for the production or modification of products excluded from patentability.

Commentary

In this provision, proposals are offered for excluding various types of biological processes from patentability.

Option 1 excludes all biological processes. While to some this is an attractive option, and others find it completely unacceptable, it is certainly contrary to TRIPs. The TRIPs Agreement only allows essentially biological processes (other than non-biological and microbiological processes for the production of plants and animals) to be excluded from patentability.

Option 2 excludes biological processes for the production of plants and animals. This is not compliant with TRIPs for the reason stated in the commentary to Option 1.

Option 3 restricts Option 2 by excluding only essentially biological processes from patentability and explicitly permitting non-biological and microbiological processes for such production. This option is in line with TRIPs Article 27.3(b).

Option 4 refers back to the articles above, excluding processes for making or modifying whatever is excluded from patentability. Option 4 is looser, and so might be challenged under TRIPs. It is likely to conflict with TRIPs, although this may depend on which products are excluded from patentability.

Article 14: Exemption of methods for treatment of human beings or animals

Option 0

no provision [i.e., there is no exemption]

Option 1

Patents shall not be granted for diagnostic, therapeutic and surgical methods for the treatment of human beings or animals.

Commentary

Option 1 is taken directly from TRIPs Article 27.3(a).

Part Two: Conditions for the grant of protection

Part Two deals with the conditions for the grant of rights. Some of those proposed are standard in patent systems (for example, novelty and inventiveness); others are specially adapted to the problems of protecting biological material.

Article 15: Conditions of protection

Paragraph 1: General requirement

Option 1

Patent protection shall be granted for inventions, whether products or processes, that are new, involve an inventive step and are capable of industrial application.

Paragraph 2: New uses

Option 1

Patent protection shall also be granted for new uses of known products or processes, provided the uses are new, involve an inventive step and are capable of industrial application.

Option 2

Patent protection shall not be granted for new uses of known products or processes.

Commentary

Paragraph 1 (General requirement): These are the customary requirements (specified in TRIPs) of novelty, non-obviousness, and utility. Additional protection requirements, e.g., prior informed consent (PIC), are set out in the following articles.

Paragraph 2 (New uses), Options 1 and 2 address the important question of new uses of known products or processes. If someone finds a formerly unknown medicinal use of a plant, or a second use of a medicinal product or of a gene, the question may arise whether a patent should be granted for this new use, as is currently the case under many patent laws. The TRIPs Agreement does not refer specifically to 'use' patents. However, even if 'use' inventions are not specifically mentioned, 'processes' are mentioned (Article 27.1), and it may be argued that 'use' claims are no more than claims to 'processes of use'. Of course, 'therapeutic processes for the treatment of humans' (which are under question here) need not be patented under TRIPs Article 27.3(a).

Article 16: Novelty

Paragraph 1: National/absolute novelty

An invention is new if it has not been:

Option 1

made available to the public in any way within [the implementing country], regardless of its existence in other countries.

Option 2

the subject of printed publication anywhere, has not been made public in any way within [the implementing country], has not been used commercially and has not been put on sale.

Option 3

made available to the public by non-confidential disclosure of any kind (whether printed, orally or by use) anywhere in the world.

Paragraph 2: Novelty of substances

Option 1

An invention consisting of biological material is not novel if the material previously existed in nature or formed part of material that previously existed in nature.

Option 2

An invention consisting of naturally occurring biological material is new if its existence was unknown prior to the application filing date.

Commentary

This provision addresses the general definition of novelty (Paragraph 1) and specific problems this requirement may pose to the patenting of inventions relating to naturally occurring substances (Paragraph 2).

Paragraph 1 (National/absolute novelty): No patent system allows the protection of what is already known — i.e., what is not novel. However, novelty can be defined in different ways.

Paragraph 1, Option 1, sometimes called 'national novelty', takes account only of what is known in the country in question.

Paragraph 1, Option 2 extends Option 1 to include printed publication everywhere.

Paragraph 1, Option 3, sometimes called 'absolute novelty', is broader again and takes account of everything previously made publicly known, whether in print, orally or by use.

All these options are TRIPs-compliant; each is based on existing models. Option 1 was formerly common (for example, in the UK and British Commonwealth countries) but is now outdated. Under this option, a company could obtain a patent in country A for a drug already in widespread use in country B, provided that information about this had not reached country A. It is doubtful how countries could benefit from this definition: it may be objected to as a way of obtaining rights to an invention already made elsewhere. Option 2 is similar to the US provisions (though the latter are somewhat more complex). Option 3 is derived from the European Patent Convention, and takes account of all public knowledge everywhere.

Paragraph 2 (Novelty of substances): These options supplement the novelty provisions. They are another way of making 'discoveries' unpatentable (compare Article 3, above).

Paragraph 2, Option 1 excludes natural material already existing. This includes material that originally formed part of naturally occurring material. This definition of novelty is harmonious with the standards set in TRIPs. Few patent laws allow protection of what already exists in nature.

Paragraph 2, Option 2 excludes only those naturally occurring biological materials whose existence was known before the application filing date. This option reflects the current patent practice of most industrialized countries (compare Guidelines for Examination in the European Patent Office, C.IV.2.3).

Article 17: Non-obviousness/inventive step

An invention is obvious (involves no inventive step) where:

Element 1
it is obvious to a person skilled in the relevant art

Element 2
it consists of natural genetic sequences, cloned natural materials or synthetic copies of biological materials existing in nature

Commentary

The terms 'inventive step' and 'non-obviousness', used in different jurisdictions, are synonymous. If every routine technical advance could be patented, innovation would come to a halt. A requirement for granting patent rights is that the invention must go beyond the routine — it must be 'inventive' or 'non-obvious'. This requirement is easy to state and very difficult to define.

Element 1 is a standard formulation of the requirement for 'inventiveness'. Element 2 is an optional addition to it, which specifically adapts the requirement to biological material.

Element 1 is TRIPs-compliant. It might be stated in different words. For example, a leading Canadian case formulates the test for obviousness (lack of invention) as follows: "would an unimaginative, skilled technician at the date of invention ... in light of his general knowledge and the literature and information on the subject available to him on that date, have been led directly and without difficulty to the invention?"[32] 'Is inventiveness present or not?' is a question ultimately judged by courts — who often disagree. Some courts — and the countries in which they are located — seem to have higher standards than others. These standards are not easy to clarify by way of legislation. For example, it would not be contributing a great deal of clarity if a legislative enactment were to state that the common-law standard in Canada of the "unimaginative, skilled technician" should be changed to that of a technician of *average* imagination.

Element 2 has two possible interpretations. It may be simply what it looks like — an adaptation of the general requirements of non-obviousness to the specific circumstances of biology. In that case, it has little point. Or it may go much farther. A frequent objection against patents on isolated gene sequences is that gene sequences are obtained by purely routine procedures, using processes that are in no way inventive. Those who make this objection may contend that Element 2 will prevent the grant of patents on such isolated sequences. However, the product is being patented, not the process. Routine processes are often used to produce non-routine and non-obvious products (a novel design of printed circuit may be made by a standard circuit-etching process; new imaginative books are written on standard word processors). If a gene and its function are known, if the isolated gene would have an obvious use, and if it is straightforward to isolate it by known means, then there is a strong argument that the isolated gene is obvious. However, the routine nature of the isolation process by itself is not conclusive. If Element 2 is interpreted as disallowing patents on non-obvious products made by obvious processes, it may be attacked as not conforming to TRIPs. It is clear that TRIPs does not define 'non-obviousness': but this does not give any licence to member countries to define the term in any fashion that they choose.

Article 18: Utility/industrial application

Option 1

An invention is useful / susceptible of industrial application if it can be made or used in any kind of industry, including agriculture.

Option 2

An invention is useful / susceptible of industrial application if it can be made or used in any kind of industry, including agriculture, and is socially useful.

Commentary

Knowledge in itself is not patentable even if it is novel and non-obvious. If the knowledge cannot be packaged in some practical (technical or industrial) application, it cannot be patented. Most patent laws require inventions to be either 'useful' or 'capable of industrial application'; the two terms are usually considered synonymous, and under their current interpretation they exclude very few inventions from patentability. Any activity that belongs to the useful or practical arts as distinct from the aesthetic arts is held to be capable of industrial application under the EPC (compare Guidelines for Examination in the European Patent Office, C.IV.4.1). It should be mentioned, however, that (mainly for historical reasons) the patent laws of quite a few countries rule that methods for treatment of the human and animal bodies shall not be regarded as inventions capable of industrial application. This exemption is addressed in Article 14.

Option 1 states that inventions must be subject to industrial or technical application. Option 2 requires inventions to be socially beneficial as well.

Option 1 states the requirement for utility very broadly. Agriculture is an industry, but art and literature are not, so a novel or painting is not patentable (though a new pigment or method of printing could be). If an invention does not work at all, it is not useful. Usually patent offices cannot test this, but they would reject, for example, an application relating to a process alleged to operate in a manner clearly contrary to well-established physical laws, e.g., a perpetual motion machine.

Option 2 is a less usual provision, and open to challenge under TRIPs. In Canada and most other countries (the USSR was once an exception), 'useful' takes no account of the social benefit or detriment of an invention. For example, in the *Visx* case (Canada),[33] it was not considered a bar to usefulness that the applicant's ray gun was designed to kill people. Social utility is important, but also extremely challenging to judge, and it is difficult to decide who should take the decision. Likewise, the definition of 'social utility' as such is elusive, since some innovations may be useful for some parts of the society and detrimental to others. Different positions among Crucible Group members are set out in Viewpoint box 17.

Article 19: Additional protection requirements

Patent protection shall not be provided unless:

Element 1

the applicant provides a certificate of origin (CO) regarding the biological material he or she relied upon in the course of developing the invention [and]

Element 2

the applicant obtains the prior informed consent of the [individuals], [institutions] [and] [or] [communities] from which he or she obtained biological materials relied upon in the course of developing the invention, provided the relevant legislation of the country of origin requires such a consent.

Commentary

This article offers a selection of conditions that inventions may be required to meet in addition to those mentioned earlier.

Element 1 proposes that patent applications provide a CO wherever biological material has been involved in the development of the invention.

Element 2 proposes that patent protection not be granted without the PIC of the provider of biological material.

The same requirements were considered previously in the options for laws governing plant variety protection (Topic 3, Section 2, Articles 12 and 13).

It has been argued that these requirements might conflict with TRIPs if they are included in a national patent or plant variety protection law as conditions for protection. On the other hand, they might not conflict with TRIPs if they are included as administrative requirements. They could be introduced in the technical examination stage. Clearly this strategy would not be foolproof, inasmuch as it could still be challenged as constituting an additional protection criterion. For an example of this approach, see the Colombian draft access legislation, Article 10(d)(h) of Colombian decree 533, 8 March 1994, 'Filing the application and acceptance or rejection thereof', wherein CO, not PIC, is introduced.

Attempts to include such requirements in the Patent Law Treaty (June 2000, on harmonization of formal requirements for patent applicants) nearly led to the failure of the Treaty. Ultimately, they were not included.

Of course, they could be included in a national patent law as long as they applied only to subject matter that may be excluded from patentability under TRIPs (e.g., plants and animals).

Another possible approach to integrating these requirements into a national patent or plant variety protection law would be to de-link them from patent validity (i.e., satisfying the CO and PIC requirements would not be a necessary precondition of the patent grant) and to link them to some other penalties. Breach of these requirements could instead give rise to a fine or a requirement to file annual accounts of efforts to trace the original source of the material and/or obtain PIC.

Viewpoint box 17: Should plant patent grants be subject to 'social utility' criteria?

Social utility tests would probably be considered a violation of TRIPs.

After the year 2000, GATT 1994 Article 23.1(b)(c) and TRIPs Article 64.2 will make it possible to bring actions to the WTO dispute settlement body for alleged violations of the *spirit* of TRIPs, independently of the fact that alleged offences do not violate standards explicitly set out in TRIPs. These actions have a great deal of potential to further restrict the range of initiatives policy-makers could introduce into patent laws. Following this line of reasoning, additional social-usefulness criteria that would operate to exclude otherwise protectable subject matter might violate the spirit of the agreement.

Let the market decide.

A 'social utility' or 'public benefit' test would open a Pandora's box that could not possibly be resolved through law. It's not easy to decide — let alone obtain wide agreement on — what is socially useful. (Try it yourself: cars? television? Internet? DDT? 'Green Revolution' rice varieties? contraceptive pills?) Moreover, views change quite rapidly over the 20-year life of a patent. A decision is certainly not within the competence of a single patent examiner. Patenting everything first, and then letting the market determine what is, and what is not, worth having is value-neutral and ultimately the most sensible way to deal with the issue. Besides, refusing a patent does not stop people from exploiting the patented invention. If something is truly objectionable, it should be outlawed, not just refused patent protection. Morally objectionable research and development should be dealt with through other systems, such as criminal law.

Property rights are government sanctions: the public has a role to play in setting priorities.

Markets normally determine private usefulness, not social usefulness. Social usefulness deals with externalities that are not captured by market prices, and granting patent protection is not a value-neutral act. Patents are state-sanctioned benefits. The government has a responsibility, when bestowing benefits, to make sure that the public's interests are represented. Patents are said to be intended as incentives for intellectual efforts, for financial investments and as a reward for the disclosure of knowledge that would otherwise be kept secret. Patent law should, therefore, not reward inventions that are not beneficial to society. Given the controversial content of many biotech-related innovations, patent legislation must have some legal mechanism whereby the public interest in the technology is considered.

Viewpoint box 18: Should certificates of origin (CO) and prior informed consent (PIC) be included as conditions for patents?

They should be considered minimum requirements for fair patents.

If there are to be patents on biological materials, as TRIPs demands, it does not seem unfair, at a minimum, to require patent applicants to disclose where they obtained the materials they used in their innovations, and that they got permission to use them.

It is in conflict with the spirit of the CBD to grant patents relating to genetic materials, regardless of the legal status of the resources the genetic materials have been derived or isolated from.

One can put together a whole set of funny situations in which thousands of legally complex questions would arise. Alternatively, one could just look for pragmatic solutions and address some of the most important issues in a way that will not require cumbersome bureaucratic procedures. No doubt there are many difficulties to be overcome when PIC and

They won't work! They aren't practical!

First of all, to the extent that WTO members must grant patent protection for biological material such as microorganisms, the CO and PIC requirements are not TRIPs-compatible, because TRIPs exhaustively lists the conditions under which patents must be made available for those materials (invention, novelty, inventiveness, utility).

Second, there is no necessary conflict between TRIPs and the CBD. They are dealing with different topics and have different emphases. Alleged instances of conflict are generally due to misunderstandings. Most countries are members of both conventions and must respect both equally.

Moreover, there are so many problems with requiring PIC and CO as conditions for the grant of intellectual property rights that it is difficult to know where to start. In fact, that is the first problem: where do such rights start? Do they start at the date the CBD came into force (December 1993), or do they go back earlier? There are understandable demands for an earlier date, but how much earlier? Any fixed date will be arbitrary, while to go back indefinitely is not practical, and, furthermore, leads to questions about the continuity of nations and past political groupings. (Would the UK have a claim to share in any benefits from the macadamia nut because its land of origin — Queensland, Australia — was once a British possession?) Then again, most countries have not legislated the conditions under which PIC and CO will be regulated and who will be involved. Until such legislation is widely in place (which may take decades), PIC and CO will not be generally available. When it is, it is probable that bureaucratic procedures will materially hinder the process of innovation, even where the source of a material is clear and not in controversy.

Another problem is the nature of PIC. What kind of information must the party obtaining PIC provide in order for the suppliers' consent to be informed? The fact is, sample collectors may not have a very clear idea of what they want to do with the material when they are collecting it, making it hard to satisfy suppliers about the role in future innovations of the material they are supplying. Or collectors may find out only after collecting the material that it is useful in some unforeseen innovation. Clear rules are needed for a workable system.

There is a specific problem with PIC and CO for patented inventions. Inventions make different types of uses of biological material. First, there are the specific biological materials that have actually been used, and the broader general class of biological materials that might be used. Inventions, unlike plant varieties, are usually generic, not specific. One

CO are made patent requirements. However, none of these hurdles seems insurmountable. Of course, the patent law would have to specify the conditions under which PIC must be given, and it would have to clarify that applicants need PIC only for the use of material actually involved in the development of the claimed innovation.

cannot obtain PIC for all members of a class of materials — the number of members is unlimited. Then there is the question of the extent to which the nature of the particular biological material is crucial to the invention. Generally, in discussion of CO and PIC, it is assumed that there is one unique sample, of known origin, the use of which is crucial to the invention. This situation happens, but it is by no means the only situation, and probably not the most common. By way of example, consider the following fact situation (hypothetical, but hardly exaggerated):

Invention: A (specified) anti-sense DNA-ripening gene driven by (any suitable) constitutive promoter, used to delay ripening in fruit and vegetables. The specification shows several specific examples, and suggests many alternatives and uses. The ripening gene was originally obtained from a UK apple variety, although it is found in one form or another in most fruit species. One of the suitable constitutive promoters (used in several examples) was obtained from cucumber mosaic virus, which is endemic in nearly all countries that grow cucumbers. No one can establish the original source of the particular promoter, which has been circulating widely in academic circles for some years. The specification gives detailed working examples of transformed apples (two varieties, one British and one Mexican), melons (one US and one Spanish variety) and bananas ('bought in a UK supermarket'), and proposes and claims (without giving any experimental detail) use of the constructs in peaches, guavas and durian.

Question: From whom, and for what, should CO and PIC be obtained in this case?

Because of all these problems, the demand for CO and PIC is seen by some simply as a way to deny patent protection to many meritorious biological inventions.

The mechanics and criteria for obtaining PIC in the context of a national bio-access law are discussed in this volume under Topic 1, Section 2, Articles 6–13.

Part Three: Rights conferred

Introduction

Part Three begins with options regarding who may be entitled to the patent. Subsequently, we address the question which rights the rights-holders under this law shall have and which they shall not have. With regard to the rights conferred, we mainly draw on TRIPs Article 28. However, with regard to those items that may be excluded from patentability under TRIPs, in particular plants and animals, WTO members are free to grant rights different from the rights they must grant for inventions that cannot be excluded from patentability. Therefore, a WTO member could decide to offer patent protection for new animal breeds, but could design the associated rights to be narrower than the rights conferred by 'normal' patents.

Article 20: Entitlement to the right

Paragraph 1: Employee-inventors

Option 1

The right to the patent shall belong to the inventors or their successors in title. If the inventor is an employee, the right to the patent shall be determined in accordance with the law of the state in which the employee is mainly employed. If the state in which the employee is mainly employed cannot be determined, the applicable law shall be that of the state in which the employer has the place of business to which the employee is attached.

Paragraph 2: Multiple inventors

Option 1

If two or more persons have made an invention independently of each other, the right to the patent shall belong to the person whose patent application has the earliest filing date.

Option 2

If two or more persons have made an invention independently of each other, the right to the patent shall belong to the person who made the invention first.

Commentary

Paragraph 1 (Employee-inventors): The right to the invention usually belongs to the inventor. The situation is more difficult if the inventor is an employee, or if several inventors have made the same invention. Different solutions are possible for the first scenario, where the inventor is an employee. The solution offered here refers to the law of the state in which the employee is mainly employed or the law of the state in which the employer has his or her place of business to which the employee is attached.

Paragraph 2 (Multiple Inventors), Option 1 reflects the first-to-file principle.

Paragraph 2, Option 2 reflects the first-to-invent principle.

Both options have advantages and disadvantages. In theory, the first-to-invent principle seems fairer than the first-to-file principle. On the other hand, it means that legislators have to define what constitutes 'making an invention' and on what date it is complete (not as easy as it sounds), and inventors must prove that they actually made the invention first, which may be extremely difficult and expensive.

Article 21: Acts requiring the rights-holder's authorization

Paragraph 1: Products

Where the subject matter of a patent is a product, the patent confers on the patent-holder the right to prevent third parties who have not obtained the rights-holder's consent from:

Element 1

making that product

Element 2

using that product

Element 3

offering that product for sale

Element 4

selling that product

Element 5

importing that product for the purposes above

Paragraph 2: Biological materials derived from patented products

The protection conferred by a patent on a biological material possessing specific characteristics as a result of the invention:

Option 1

shall extend to any biological material derived from that biological material through propagation or multiplication in an identical or divergent form and possessing those same characteristics.

Option 2

shall not extend to biological material that has merely been derived from that biological material through propagation or multiplication.

Paragraph 3: Processes

Where the subject matter of a patent is a process, the patent confers on the patent-holder the right to prevent third parties who have not obtained the owner's consent from:

Element 1

using the process

Element 2
using the product obtained directly by the process

Element 3
offering for sale the product obtained directly by the process

Element 4
selling the product obtained directly by the process

Element 5
importing for use or sale the product obtained directly by the process

Paragraph 4: Biological materials derived from patented processes
The protection conferred by a patent on a process that enables a biological material possessing specific characteristics to be produced as a result of the invention shall extend to biological material directly obtained through that process and:

Option 1
to any other biological material derived from the directly obtained biological material through propagation or multiplication in an identical or divergent form and possessing those same characteristics.

Option 2
shall not extend to further biological material obtained through propagation or multiplication of such directly obtained biological material.

Commentary
Paragraphs 1 and 3 build on TRIPs Article 28. As mentioned in the introduction to this part, there is no need for WTO members to follow Article 28 with regard to those inventions that may be excluded from patentability under TRIPs. For those inventions, WTO members may grant rights different from the rights foreseen in TRIPs.

Paragraph 1 (Products) deals with product patents.

Paragraph 2 (Biological materials derived from patented products), Options 1 and 2 raise the question of what a patent on biological material actually signifies. Does it mean that the patent-holder has the right to prevent third parties from using any biological material derived from the patented material through propagation or multiplication in an identical or divergent form and which possesses the same characteristics as the patented material (Option 1)? Or does the patent only give the patent-holder more limited exclusive rights, not including, for example, rights over progeny or derivatives (Option 2)? The TRIPs Agreement does not address directly what rights a patent over biological material must give. However, some might argue that a product patent that does not protect against competition from identical products obtained by multiplication of the original does not meet the requirements of TRIPs Article 28.1(a).

Paragraph 3 (Processes) deals with the scope of process patents. Under TRIPs, process patents shall confer on the patent-holder the right to prevent third parties who have not obtained the owner's consent from using the patented process. In addition, patent-holders shall have the right to prevent others from the acts of using, offering for sale, selling or importing for these purposes the product obtained directly by that process. To be TRIPs-compliant, therefore, laws should include all of Elements 1–5.

Paragraph 4 (Biological materials derived from patented processes): With regard to patents on processes for the modification of biological material, the question arises whether the patent should extend only to the biological material directly obtained through that process (Option 2) or also to biological material derived through propagation or multiplication from the directly obtained biological material (Option 1).

Article 22: Exhaustion of rights

Paragraph 1: Product patents

Option 1

The protection conferred by a patent on a biological material shall not extend to biological material obtained from the propagation or multiplication of material placed on the market [in the territory of this country] by or with the consent of the holder of the patent, where the multiplication or propagation necessarily results from the application for which the biological material was marketed, provided that the material obtained is not subsequently used for other propagation or multiplication.

Option 2

The protection conferred by a patent on a biological material shall not extend to biological material obtained from the propagation or multiplication of material placed on the market [in the territory of this country] by or with the consent of the holder of the patent.

Paragraph 2: Products directly obtained

Option 1

The protection conferred by a patent on a process that enables a biological material possessing specific characteristics to be produced as a result of the invention shall not extend to biological material obtained from the propagation or multiplication of biological material placed on the market [in the territory of this country] by or with consent of the holder of the patent, where the multiplication or propagation necessarily results from the application for which the biological material was marketed, provided that the material obtained is not subsequently used for other propagation or multiplication.

Option 2

The protection conferred by a patent on a process that enables a biological material possessing specific characteristics to be produced as a result of the invention shall not extend to biological material obtained from the propagation or multiplication of biological material placed on the market [in the territory of this country] by or with consent of the holder of the patent.

Commentary

This provision deals with the exhaustion of patents. If someone buys a patented product from the patentee, they expect to be able to use and sell it. A patent is said to be 'exhausted' if the patented invention has been placed on the market by (or with the consent of) the patent-holder. If someone then buys the patented product, the use of this product is free — the patent-holder's rights do not extend to this use any more. The patent-holder's exclusive right remains to reproduce the product. In the case of patented self-replicating material marketed with the patent-holder's consent, the question arises whether any further propagation of that material requires the rights-holder's consent.

Paragraph 1 (Product patents) deals with the exhaustion of product patents.

Paragraph 1, Option 1 is relevant only if product patents actually extend to biological material derived from patented biological material through propagation or multiplication (Article 21, Paragraph 2, Option 1). If this were the case, a patent granted for plant seeds would also extend to the harvest produced from the seeds. Consequently, those who wish to sell or use the harvested product, for example for the production of flour, would require the patent-holder's consent. In order to avoid this, Option 1 provides (notwithstanding Article 21, Paragraph 2, Option 1) that the patent shall not extend to the harvest obtained from the propagation of the patented seed, provided that the seeds were placed on the market by the patent-holder for the purpose of propagation. The farmer, having bought seeds, expects at least to be able to plant them, harvest them, and sell them for consumption. If, however, the farmer wishes to use or sell harvested material for further propagation, the patent would not be held to be exhausted. This option gives the holder of a patent control similar to that of the holder of plant variety rights; however, if the farmers' exemption (see Article 23, below) is recognized for plant variety protection and not patents, this can make a significant difference.

Paragraph 1, Option 2 takes a more radical approach and is attractive only to those countries that wish to minimize patent protection in this field while arguably complying with TRIPs. The patent would be exhausted as soon as seed is sold for the purpose of propagation, in respect of that seed and all its progeny. Thus, farmers could freely replant harvest produced from that seed without the patent-holder's consent. Further, farmers and others could freely multiply and sell such seed in competition with the patentee. Option 2 gives the patentee very limited rights: in effect, the right to sell patented seed free of competition for a single season only. This certainly will not be regarded as satisfactory by patentees, and could be challenged as contrary to TRIPs Article 30 (as constituting unreasonable prejudice to the legitimate interests of the patent owner).

Paragraph 2 (Products directly obtained) deals with the exhaustion of process patents.

Paragraph 2, Option 1 is relevant only if the protection conferred by a process patent also extends to biological material derived from the directly obtained biological material through propagation or multiplication (Article 21, Paragraph 4, Option 1). If this were the case, the process patent would also extend to the harvest produced from the patented material. Consequently, those who wish to sell or use the harvested product would again require the patent-holder's consent. To avoid this, Option 1 provides (notwithstanding Article 21, Paragraph 4, Option 1) that the process patent shall not extend to the harvest obtained from the propagation of the patented seed, provided that the seeds were placed on the market by the patent-holder for the purpose of propagation. If, however, the farmer wishes to use harvested material for propagating purposes, the patent would not be held to be exhausted.

Paragraph 2, Option 2 takes a more radical approach and is attractive only to those countries that wish to minimize patent protection in this field while arguably complying with TRIPs. The patent on the seeds derived from material that has been modified with a patented process would be exhausted as soon as it has been sold for the purpose of propagation. Thus, farmers could freely replant harvest produced from that seed without the patent-holder's consent, and farmers and others could freely multiply and sell such seed in competition with the patentee. Option 2 is consistent with Article 21, Paragraph 4, Option 2 (although Option 2 here is only concerned with seed sold by the patentee, whereas Article 21, Paragraph 4, Option 2 is concerned with all seeds).

Article 23: Exemptions from the rights conferred

Paragraph 1: Private use

Option 0

no provision

Option 1

The rights conferred on patentees shall not apply to activities done privately and for non-commercial purposes.

Paragraph 2: Research

Option 0

no provision

Option 1

The rights conferred shall not apply to acts done for experimental purposes.

Option 2

The rights conferred shall not apply to acts that constitute research relating to the patented inventions.

Paragraph 3: Breeding

Option 0

no provision

Option 1

Acts done for the purpose and in the course of breeding and developing new varieties shall not constitute infringement.

Paragraph 4: Farming

Option 0

no provision

Option 1

The responsible minister may, within reasonable limits and subject to the safeguarding of the legitimate interests of breeders, restrict Breeders' Rights in relation to varieties of specified plant genera and species in order to permit [small-scale] farmers to use for propagating purposes, on their own land, the product of the harvest which they have obtained by planting, on their own land, a variety of any plant genus or species that is the subject of any rights conferred under this Law.

Option 2

Notwithstanding the rights conferred under this Law, [small-scale] farmers shall be allowed to plant-back, on their own land, seeds that are grown on their own land and to exchange seeds with other farmers on a non-commercial basis.

Option 3

Notwithstanding the rights conferred under this Law, [small-scale] farmers shall be allowed to plant-back, on their own land, seeds that are grown on their own land, to exchange seeds with other farmers on a non-commercial basis, and to sell seeds in the following limited quantities [provide details of quantities].

Paragraph 5: Local derivation

Option 0

no provision

Option 1

Residents of this country shall be allowed to use biological inventions derived from or based on germplasm collected in [the implementing country] without the need to seek permission from the rights-holder [on payment of royalty at a rate to be determined by government regulation].

Commentary

All patent legislation recognizes a variety of exemptions to granted rights. Here we list several elements that could be excluded.

Paragraph 1 addresses purely private acts, provided they are done for non-commercial ends. Paragraph 2 presents various proposals for a research exemption. Paragraph 3 addresses an exemption for breeders. Paragraph 4 includes various exemptions concerning farmers. Paragraph 5 presents an exemption for inventions that have been made on the basis of material of local origin, and for customary uses.

Options for the definition of 'small-scale farmers' are presented in Topic 3, Section 2, Article 6, above.

Paragraph 1 (Private use): The private-use exemption is very common. In some laws it is implicit, in others explicit. It conforms with TRIPs Article 30, as it in no way interferes with the patentee's normal enjoyment of the rights conferred by the patent.

Paragraph 2 (Research), Option 0 rejects the idea of a research exemption. This provision may not appear in any existing law, but is not too far from representing US case law. In the United States (apart from a specific '*Bolar*' exemption for experimentation with patented drugs to meet safety regulation requirements), the exemption is only for 'philosophical inquiry' — seeking knowledge for its own sake. Firms never do this, and academics less often than formerly. The disadvantage of this option is that it may disproportionately hamper research on biological materials, which (even when obtained from the patentee) can infringe when reproduced. For example, in the United States, the owner of a patented variety can prohibit any use by a competitor, for any purpose, of a patented plant variety or plant part. This prohibition even extends to the use of patent-protected plant parts in research that culminates in the creation of a plant variety that did not include the protected material.

Paragraph 2, Option 3 is the form of research exemption common in Europe. It allows experiments on the patented invention (to see how it works, or improve it) but not regular use of the invention as part of commercial development of a product. It is not clear, however, to what extent patented plant material could be used, without the patent-holder's consent, for the development of other material.

Paragraph 3 (Breeding) introduces a special exemption for breeders. It may be useful to specify the usual research exemption with regard to breeding activities. Note that the breeding exemption does not exempt commercial exploitation of products of the breeding activity. Whether such products infringe must be judged by comparison of the resulting varieties with relevant rights, taking into account applicable provisions under Article 21, Paragraphs 1 and 2.

Viewpoint box 19: How do TRIPs principles apply to exemptions?

Extra-broad exemptions are possible for patents over subject matter not covered by TRIPs.

To comply with TRIPs, the exemptions from patent rights in national legislation can be as broad as one wishes as long as patenting the subject matter is not mandated by TRIPs. Consequently, national legislation can have very wide exemptions for plants (but not necessarily plant varieties), animals and essentially biological processes. This follows from Article 27.3, which allows members to exempt these categories from patentability and thus from the applicability of the TRIPs principles as laid down in Article 27.1.

Thus, even where a member decides to grant, for example, patents for traditionally bred but not for transgenic animals, this member would not violate the TRIPs principle according to which patents shall be available 'in all fields of technology'.

Once a WTO member state offers patent protection, all exemptions must comply with TRIPs, even for subject matter not covered by TRIPs.

TRIPs requires patents to be granted without discrimination as to subject matter. For matter that does not have to be protected, countries have the right to refuse patents; however, should they decide to grant such patents, they then have the obligation not to discriminate according to subject matter, and must grant the same rights for all patents.

Recommendation

The Crucible Group recommends that any systems for protection for plant varieties should provide 'appropriate' exemptions for farming and breeding activities. The breeding exemption should allow, as a minimum, research on protected material and the development of products from it to fall outside the scope of the rights (in the case of patents, outside the definition given in the patent claims). The scope of such exemptions should be decided in the light of specific circumstances.

Paragraph 4 (Farming) introduces the farmers' exemption into patent law. The options listed are identical with those set out above in Topic 3, Section 2, Article 17, Paragraph 3. Option 2, giving farmers the right to deal commercially in protected seed of plant varieties in competition with the patentee, might be totally unacceptable to most formal breeders, who would certainly argue that such rights for farmers would not satisfy the TRIPs requirement to establish an 'effective *sui generis* system' for the protection of plant varieties.

Paragraph 5 (Local derivation): Option 1 provides an exemption for use, by nationals, of biological material of local origin. For materials of mixed origin (e.g., obtained by cross-breeding) one might need to specify what proportion of national origin would be sufficient. The preference for local residents conflicts with the national treatment principle of TRIPs if the material in question may not be exempted from patentability. Moreover, Option 1 would conflict with TRIPs because it provides an automatic compulsory licence.

Article 24: Derogation from exemptions

Option 0

no provision

Option 1

Notwithstanding any private contracts restricting the use of inventions, the exemptions as set out in Article 23 shall apply.

Commentary

The purpose of this article is to prevent patent-holders and parties who would otherwise take advantage of the exemptions set out in an intellectual property law from 'contracting out' those exemptions (in other words, creating restrictions on other parties' uses of materials that would otherwise be allowable pursuant to exemptions). Some members of the Crucible Group feel that the inclusion of a clause such as this is essential, as the exemptions reflect important public policy positions; consequently, patent owners should not be allowed to undermine exemptions through private negotiated deals. Other members feel that such a clause represents unnecessary interference with parties' freedom to make whatever private contractual deals they want.

Article 25: Duration

Patent rights shall last for 20 [or other number] years from the application date.

Commentary

A term of patent protection must be prescribed in the patent law. To conform with TRIPs Article 33, this term should be not less than 20 years from the date of the patent application.

Viewpoint box 20: Should duration of patent rights vary for different subject matter?

The duration of the rights should always be the same.

Setting the term of intellectual property rights is not an exact science. Some inventions need or justify longer protection. However, there is no agreed way of deciding the term, and it is impractical to do it case by case. The conclusion of the TRIPs negotiators was that a term of 20 years sufficed in most cases. This is a reasonable practical compromise for general use.

Durations should be different for different classes of things.

There may be problems associated with discriminating among different subject matter, but there are competing principles to consider.

Options regarding what rights to confer upon patent-holders may vary depending upon the combination of (a) the strictness of criteria for protection, and (b) the breadth of subject matter that can be protected pursuant to those criteria. Stronger rights, such as a longer term, the right to exclude others from activities like duplication, selling, importing, and so on, should be reserved for patents that extend to a relatively narrow range of subject matter.

Article 26: Compulsory licences

Paragraph 1: Formal conditions

A compulsory licence may be granted under any patent on a biological material or process to an applicant who shows:

Element 1

the ability to work the patented invention

Element 2
that the patent has been granted for at least three years

Element 3
that the patentee has refused to grant a licence on reasonable terms (or at all)

Paragraph 2: Substantial reasons
No compulsory licence may be granted unless it is in the public interest that the applicant be granted a licence to work the patent in the country for one or more of the following reasons:

Element 1
the patentee is not exploiting the invention in the country

Element 2
the patentee is supplying the market for the invention in the country primarily by importation

Element 3
the patentee is not supplying a market or potential market for the invention in the country on reasonable terms

Element 4
the applicant has made a significant improvement to the patentee's invention, which the applicant is unable to exploit in the country by reason of the patentee's refusal to license on reasonable terms

Element 5
the grant of a licence to the applicant would promote public health, food security or the protection of the environment within the territory

Element 6
the patentee has followed anticompetitive practices with respect to the subject matter of the invention

Element 7
the patented invention makes use of or is derived from biological material originating in the country

Commentary
Paragraphs 1 and 2 set out some elements that might be included in a compulsory licence provision. It is important, while reading through these options, to remember that TRIPs Article 31 sets out a list of relatively restrictive conditions for the grant of compulsory licenses. The following comments provide details

regarding the relationship of the elements provided in this article to those included in Article 31 of TRIPs.

Paragraph 1 (Formal conditions) sets out the requirements to be met by the applicant for the licence. Any element may be omitted from the law; however, any element included must be satisfied by the applicant. To comply with TRIPs, Elements 2 and 3 must be included: Element 2 is specified in the Paris Convention, while Element 3 is the subject of TRIPs Article 30(b).

Paragraph 2 (Substantial reasons) sets out possible grounds for grant of a compulsory licence; the applicant need establish only one of the grounds included. Some of these grounds may not be TRIPs-compliant. Element 2 does not comply with TRIPs because Article 27.1 does not allow discrimination against exploitation by importation. Element 7 may be contrary to TRIPs because it discriminates according to country of origin of the material patented, arguably infringing Article 27.1. Element 5 could be argued to discriminate according to field of technology, or perhaps as being so broadly phrased as to fall foul of TRIPs Article 30.

Compulsory licence provisions are included in the patent laws of most countries (although not in the United States, where such licences may be granted to remedy breaches of anti-trust laws). There are two main types of compulsory licences: discretionary and automatic.

Automatic compulsory licences were formerly granted in many countries, typically for inventions relating to food or pharmaceuticals. This was done to ensure that innovations relating to food and medicine were made available at the lowest possible prices. Patentees of such inventions regarded such provisions as confiscation of their rights, and as a severe disincentive to innovation in important areas. Such licences are now rare, since they conflict with the obligations of TRIPs on non-discrimination as to subject matter.

Discretionary compulsory licences: In some countries, these are seen primarily as a safeguard against failure by the patentee to work the invention, or against attempts by the patentee to suppress technical advances that compete with the patentee's product. In other countries, the patentee is not obliged to work the patented invention, and compulsory licences are granted only as remedies for breaches of anti-trust law (for example, where the patentee makes the purchase or use of an unpatented product a condition for the grant of a licence to use a process patent).

In both cases, the grant of such licences is relatively rare. This is in part because the mere presence of such provisions (in territories where they exist) encourages patentees to negotiate reasonably. In part it is because, while voluntary licences frequently lead to constructive and developing business relationships, this is rare with compulsory licences. Understandably, patentees resent attempts to seek such licences, whether or not they are successful.

Even if a patent law did not include compulsory licence provisions, courts could still apply compulsory licences as a remedy for anti-trust abuses.

Article 27: Anticompetitive practices

Option 0

no provision

Option 1

The following shall be prohibited: all agreements between undertakings, decisions by associations of undertakings and concerted practices that may affect trade and that have as their object or effect the prevention, restriction or distortion of competition within the market, and in particular those which

a) directly or indirectly fix purchase or selling prices or any other trading conditions;
b) limit or control production, markets, technical development, or investment;
c) share markets or sources of supply;
d) apply dissimilar conditions to equivalent transactions with certain trading parties, thereby placing them at a competitive disadvantage; or
e) make the conclusion of contracts subject to acceptance by the other parties of supplementary obligations that, by their nature or according to commercial usage, have no connection with the subject of such contracts.

Any agreements or decisions prohibited pursuant to this Act shall be automatically void.

Commentary

Almost all political systems rely to some extent on markets to provide at least part of the economic needs of their people. Some systems emphasize economic freedom and the minimum of controls; some profoundly distrust markets, control them and restrict their operations to as few sectors as possible; most take an intermediate position somewhere between these two extremes. However, to rely on competition means combating anticompetitive practices that impede the efficient operation of free markets — practices such as price-fixing, market-sharing, agreements not to compete, and misuse of monopoly power. The greater the reliance on markets, the greater the need to ensure that they work as they should. This is as important as it is difficult.

The Crucible Group, members of which have a wide range of opinions about the value and efficiency of markets, nevertheless holds the view that anticompetitive activity can damage the prosperity of richer nations and the development of poorer ones, and can also threaten biological diversity and the environment.

However, it does not follow that anticompetitive activities should be dealt with specifically in a legislative project for the protection of biological inventions. It may be argued that the topic is too important for this. Certainly, its complexity is such that the Crucible Group cannot hope to provide an exhaustive series of options. Accordingly, Option 0 proposes no provision, and assumes that the question will be dealt with in separate laws. Option 1 is given by way of an example of the sort of provision that might be included in legislation on the protection of biotechnological inventions.

Recommendation

The Crucible Group recommends that countries should set up and enforce anti-trust laws to ensure fair competitive practices in the seed industry. Similar arrangements are needed at the international level to complement national law.

Part Four: Procedural/administrative matters

Article 28: Persons entitled to file applications

Option 1

A patent application may be filed by any natural or legal person or any body equivalent to a legal person by virtue of the law governing it.

Option 2

A patent application may be filed by:

Element 1

a national of this country

Element 2

an alien having residence [or a registered office] [or an establishment] in this country

Element 3

a locally resident agent duly authorized to act on behalf of an alien

Commentary

This article provides options concerning who may apply for patents.

Article 29: Remedies

Commentary

The rights set out in this collection of legislative options create a legal foundation for possible civil actions and/or criminal prosecutions for unauthorized use of the protected innovation. Remedies against parties found guilty of such actions could range from prohibitions to compensation, fines, probation and jail terms. Some form of appeal or judicial review would have to be made available for first-level determinations. The Crucible Group has refrained from delving into analysis of possible 'causes' of action, criminal offences, remedies and punishments related to the options presented in the previous parts.

Article 30: Competent authority

Option 1

The national patent office shall be responsible for the grant of rights under this Law.

Part Five: Relationship to other acts

Article 31: Relationship to plant variety protection law

Option 0

no provision

Option 1

Notwithstanding any patent rights restricting the use of material of plant varieties or parts thereof, the exemptions as set out in Article 23 shall apply.

Commentary

For further discussion of this issue, see 'The relationship between patent and other rights (e.g., plant variety protection)' in the Appendix.

Appendix

Background issues related to the legislative options

Contents

1. An introduction to access laws and their relationship to intellectual property laws

What are 'access' laws, and how do they relate to intellectual property protections for genetic resources and indigenous and local knowledge?

A. What is access law?

Genetic resources are either in the public domain or they are subject to some form of control. If they are in the public domain, anyone can collect them and use them. If they are under some form of exclusive or restrictive control, parties seeking to use them will have to get the permission of the parties controlling them. Access laws arise in the context of the latter scenario, where someone has or asserts a right of control.

As the previous paragraph implies, there are two distinct legal steps involved in the realization of an 'access' law. First, there must be a legal basis upon which a party can refuse others access to the resource in question, be it a tangible resource, like a plant, or an intangible resource such as knowledge related to the use of the plant. Second, optionally, there may be rules regarding (a) procedures that the parties seeking and supplying the resources must comply with in the negotiation of access agreements, and (b) minimum conditions to be included in those agreements. For example, these rules could create the procedural requirement for access-seeking parties to publish a notice of their application. They might include a minimum condition that the collecting party must make a deposit of collected samples with the national government.

These rules can apply to both bilateral and multilateral agreements. For example, in the context of the creation of a multilateral system of exchange, the parties could mutually agree to forfeit their right to say 'no' to proposed collections of protected resources. This is the nature of the proposed multilateral agreement for the International Undertaking on Plant Genetic Resources for Food and Agriculture (IU on PGRFA). In this case, it appears that the state parties who sign the multilateral access agreement will agree to suspend their right to say 'no' to applications for a closed list of plant species.

The second set of rules regarding procedures and minimum terms would not be possible without the pre-existence of parties' ability to refuse the collection of the resource in the first place. Parties could not meaningfully put themselves into a position to negotiate and possibly agree to bilateral agreements or multilateral agreements without first having the ability to prevent the very collections that are being proposed.

A.1 Primary rights of control

The legal basis for the primary right to say 'no' is rooted in different property laws and in national sovereignty. We provide descriptions of these mechanisms in the following four paragraphs.

1 Real property law gives the supplying party the right to deny others access when the desired resource is physically located on the suppliers' real property. Real property law does not require actual ownership of the land — just a recognized right of control of physical access to the land. It is the right to prohibit trespassing on the land, which is rooted in real property law, that gives the real property rights-holder the ability to deny access. *My ability to keep you off my land is the basis of your need to enter into an agreement with me if you want access to a resource located on my land. If you can locate the resource somewhere else, you need no agreement with me.*

2 Personal property law gives the supplying party the right to deny others access when the desired resource is physically located in the supplier's personal property. In this way, personal property operates just like real property as a basis to deny access to the resource. Personal ownership of a resource can be used to deny others taking it, whether or not it is on my land. *My ability to keep you from using my personal property is the basis of your need to get my permission to use it, and to approach me with terms you hope I will find acceptable. Again, if you can find the resource somewhere else, you need no agreement with me.*

3 Intellectual property law gives the right to deny others access to a resource when the resource is the embodiment of the protected intangible property. Intellectual property ownership of a resource can be used to deny others use of the resource whether or not it is on or in someone's real or personal property. The lines between these different types of law get fuzzy sometimes, because it is possible to state that someone's intellectual property forms part of their personal property. This is not terribly relevant here. Most important is the principle that intellectual property rights in the resource provide the right to deny others access to the resource in more situations than real and personal property law. This is, of course, the motivation behind the creation of *sui generis* intellectual property protections for indigenous and local knowledge.

4 Sovereign states have the right to prohibit parties from gaining access to tangible resources located on their lands. Although sovereignty does not reduce to a mere assertion of property rights, at least by analogy, national

sovereignty vests a right of control over domestic resources that is equivalent to a real property right and a personal property right (as long as the resource is originally from inside the country). As a function of its sovereignty, a state government has the right to prohibit entry to its land and to regulate the terms under which foreign parties may collect genetic resources inside the country, so the government could require, for example, that collectors obtain permission from local communities supplying desired resources. The requirement that the community be involved in the overall consent-granting process is a function of a state's underlying sovereign right of control over the resource.

None of these four bases of control depends upon what we refer to as access law. They existed long before any so-called access laws were created. Of course, a national government could put together of package of legislation that simultaneously created both rights of control over tangible resources or related knowledge, and procedural rules for negotiating collection and supply of those resources or knowledge. It is worth noting, however, that existing access laws, so far as we know, do not include the creation of such rights of control. Instead, they are limited to the second stage of rule-making and regulation. For example, the Philippines' EO 247 does not create rights of control in the state over its genetic resources. Nor does it create novel rights of control in the local populations. Instead, it is limited to the creation of procedural rules that applicants must respect when applying for access, and minimum substantive terms that must be included in those agreements.

A.2 Secondary standard-setting for negotiating supply and collection of resources subject to primary rights of control

Where primary rights of control already exist, people seeking to use the controlled resource must approach the controlling party to ask permission to use it. If they can come to an agreement, then the controlling party will supply the resource on mutually agreed terms. These agreements are contractual in nature. That is to say, they represent an agreement among parties about how both will behave. They are agreements wherein both parties have given their PIC to terms or conditions that they have the authority to agree to. In the context of resources protected by intellectual property laws, the agreement is contractual in nature, but is called a 'licence'. In principle, parties seeking to use indigenous and local knowledge protected by *sui generis* intellectual property laws would also have to obtain a licence from the 'owners' of that knowledge in order to use it. These sorts of agreements, and the necessity to negotiate agreed-upon terms with the owners of the resources, existed long before people started talking about 'access laws'. The principle of PIC is essential to all of contract law. It is not a creation of the Convention on Biological Diversity (CBD) or of national access laws. If everyone were satisfied with how such agreements were going already, there would be no need to do anything new. By now, however, the argument is familiar that left on their own, unregulated access-seeking and access-granting parties were not striking satisfactory deals. The second step in creating an access law consists of regulating the means by which these deals are struck and establishing minimum terms that must be included in those deals.

It is possible to create, by law, additional procedures that parties to such agreements must go through before the agreement can be concluded. For example, it is possible to require that the party seeking to collect plants located on indigenous lands get the written consent of the indigenous community concerned, on a form prepared in the indigenous communities' language. Likewise, by law, the collecting party might have to appear at a public hearing to answer questions about its proposed collecting activities.

Similarly, it is possible, through law, to impose minimum conditions to be included in all mutually agreed contracts for the supply and collection of genetic resources. For example, the collecting party might be required to agree to share a percentage of the profits it makes in the future from R & D that involves the resources collected. Or it could require that the collecting party include local experts in its collecting activities. In both instances — the procedural and content-based regulations — the PIC of the supplying party will not have legally binding effect unless terms are met. These rules do not create the PIC principle; they merely regulate the conditions under which PIC can be meaningfully obtained.

These rules regarding the conduct of parties are generally what are called 'access' laws. As stated above, most access laws do not include components wherein they create underlying property (or sovereign) rights over resources. They could, but they generally do not.

In the international, bilateral context, access laws are regional agreements wherein the state members of the region agree to harmonize national access laws such as those described here. Andean Pact Decision No. 391 is an example of just such an international, bilaterally based regional agreement.

In the multilateral context, access laws are international agreements wherein the state parties create a system of unfettered (or less fettered) exchange of genetic resources located within their territories, subject to a standardized (minimal) set of conditions. The IU on PGRFA is an example of such an agreement.

Most people argue that the CBD's provisions on national sovereignty, access, benefit-sharing and technology transfers work towards the standardization of bilateral exchanges. In this way, the CBD is like the Andean Pact Decision No. 391, except it is global in application and far more general. Others argue, however, that the CBD can be interpreted as encouraging multilateral agreements such as that being negotiated in the IU on PGRFA.

B. *Sui generis* protections for indigenous and local knowledge and their relationship to access laws.

As Part A set out, parties' ability to collect and/or use resources protected by intellectual property protections is limited by virtue of rights of control vested in the owners of that intellectual property. Parties seeking to use protected intellectual property must get the PIC of intellectual property owners before they can use it. *Sui generis* intellectual property protections for indigenous and local knowledge are no exception. If such laws protect indigenous and local knowledge by vesting rights of exclusive use of protected knowledge with a particular community, parties seeking to use that knowledge will have to seek the permission of the community. In making such laws, the government has fenced-off protected indigenous and local knowledge from the public domain and restricted other parties' ability to collect and use it. In so doing, they have satisfied the first prerequisite legal condition for the creation of a national 'access' law.

Once such a right of control is established, policy-makers have a choice. They can leave parties seeking to use the protected knowledge and the owners of the protected knowledge to their own devices to approach and negotiate with each other in whatever way they want. Or they can create rules regulating how the parties must deal with each other, and insist upon minimum terms in their agreements with one another.

All countries with which we are familiar adopt the former, less regulated approach with respect to mainstream patent, copyright, design and plant variety protection schemes. For example, once the intellectual property rights of control over a plant variety are established, it is left entirely to the parties to negotiate, if desired, terms of the access, supply or use of the protected plant variety.

It is not clear if policy-makers would be so laissez-faire with respect to indigenous and local knowledge that is protected by *sui generis* intellectual property laws. For example, a government may want very much to make additional rules regarding how parties negotiate agreements (or licences) to use protected indigenous and local knowledge. These required terms could include procedural guarantees designed to protect indigenous and local communities from exploitation, such as the requirement that requests for permission be translated into the local language, be subject to public hearings, or be subject to review by national governments, etc. They could also include minimum terms of agreement designed to protect the communities' interests (such as mandatory sharing of a certain percentage of profits from downstream commercialization). Laws that include such rules would overlap into the second (optional) stage of regulating access. (Note: in Topic 2, Section 3, the Crucible Group includes such options. Part Five, Article 19 provides options for secondary (optional) rules regulating contractual relationships that flow from the underlying *sui generis* intellectual property rights established in Parts One to Four.)

On the other hand, governments could simply choose to establish the right of control over their knowledge and leave the communities to negotiate terms as communities themselves feel is appropriate. In either case, it is important to keep the potential stages in the development of both *sui generis* intellectual property protections for indigenous and local knowledge separate. It is true that access law could include the creation of both primary rights of control and secondary (optional) rules regarding contractual relations between the parties. Similarly, *sui generis* intellectual property protections for indigenous and local knowledge (or farmers' varieties, for that matter) could include not only the creation of primary rights of control, but the secondary procedural requirements and minimum terms regarding contractual relationships. Whether or not the two stages are actually included in a single law, it is important to remember that, conceptually at least, they are distinct.

2. Relationship between patent and other rights (e.g., plant variety protection)

Interface problems may arise where (1) a state decides to provide for the protection of plant varieties by a combination of patents and a *sui generis* system ('double protection'), and (2) different forms of intellectual property protection affect the use of one and the same plant variety ('overlapping protection').

A. Double protection

Article 27.3(b) TRIPs allows the protection of plant varieties by patents, by *sui generis* rights and by "any combination thereof". The 1978 UPOV Convention allows protection of new plant varieties by means of a "special title of protection" or a patent. Both forms of protection must not, however, be provided for one and the same botanical genus or species.[34] The 1991 UPOV Convention does not contain this prohibition of double protection. It only requires that adhering states protect all plant genera and species in accordance with the provisions of that Convention.

One rationale behind the ban on double protection was that problems might arise where patents and Plant Breeders' Rights are granted for varieties of one and the same species or genus. If some varieties are patented while others are protected by Plant Breeders' Rights, farmers and breeders might face a rather confusing situation: the varieties protected by Plant Breeders' Rights could be used freely for breeding purposes and replanting, while the use of the patented varieties for these purposes would require the patent-holder's authorization. States that consider such problems serious may choose to avoid them by excluding plant varieties from patentability and instead establishing a *sui generis* system for the protection of plant varieties. This approach would be in line with TRIPs as well as with UPOV 1991.

B. Overlapping protection

Patent law and variety protection law are two altogether independent forms of intellectual property protection. Although they have common principles, they are subject to different rulings in respect of the subject matter, the protection requirements and the effects of protection. In the past, when commercial plant breeding was basically in the hands of plant breeders, the two forms of intellectual property protection coexisted quite smoothly; however, modern biotechnology has increasingly enabled economic sectors outside the breeding industry to make innovations relevant for the breeding of plant varieties. Should states decide to grant patents for these biotechnological innovations, the question may arise how such patents should affect the exercise of possible privileges (breeders' and farmers' exemptions) afforded by plant variety protection.

Possible overlaps of different forms of protection are demonstrated by the following example:

A patented gene is inserted into a plant that is protected under a UPOV-type breeders' right. Can the plant be used freely under the breeders' exemption as an initial source for breeding a new variety, or would such use infringe the patent on the inserted gene? A further question is whether farmers may plant-back saved seed of that variety as allowed under the farmers' exemption, or whether the planting-back would require the patent-holder's authorization.

There are two approaches:

1 There is no interface problem and no need to make special provisions.
2 Conflict between the rights given by patents and in other statutes requires special interface provisions.

According to the first view, intellectual property rights give *only a right to exclude*. Everything that is not specifically forbidden is allowed. However, actions that are not forbidden under one statute may be forbidden under another. Each system is independent. For example, while a book may not be subject to patent rights as an artistic creation, use of it may still be forbidden as a breach of copyright; likewise, replanting of farm-saved seed may be allowed under plant variety rights, but still forbidden if the seed contains a patented gene. According to this view, any use of the plant variety for breeding purposes would also make use of the patented gene and thus require the patent-holder's authorization, at least as long as the new variety contains (and expresses) the patented gene.

In the second view, intellectual property laws create, on the one hand, rights to exclude, but recognize, on the other hand, certain positive rights which remain unaffected. In this view, plant variety protection laws may acknowledge general principles, embodied, for example, in the right of farmers to save seed (farmers' exemption) and the right of breeders to use protected varieties as an initial source for breeding new varieties (breeders' exemption). If so, these rights should not be overridden by patent claims relating to a specific gene or trait of the variety, or to a process by which the plant cell from which the variety has been generated was transformed. To determine under which circumstances the rights should be overridden requires special interface provisions.

Claimed advantages of the first approach include simplicity and encouragement of investment through strong patent rights. Rights to conduct research and to private non-commercial use ensure, it is claimed, balance and the right level of freedom. On behalf of the second approach, it is said that the special conditions of farmers, in particular in developing countries, and the need to ensure continued access to germplasm justify a special approach to intellectual property legislation with regard to biotechnological innovations.

Consequences

1 For the first approach, no special interface provisions are necessary. However, appropriate general exemptions for research and private use, applicable to all (not just farmers), must be in place. Such exemptions may typically result in freedom for subsistence farmers to save and reuse seed containing patented genes, though not to sell their harvests commercially.

2 For the second approach, a decision is required about where to put the interface so as to balance the rights of patentees and others effectively. One way to avoid some interface problems is to exclude innovations that make use of or relate to genetic material from patentability. In addition, or alternatively, one may address interface problems by special rules on exhaustion of patent rights on genetic material. Options include:
   a) Naturally occurring substances, including genetic materials, are excluded from patentability. As it may be argued that those substances neither constitute an invention nor satisfy the novelty requirement, their exclusion would — arguably — be compatible with TRIPs.[35]
   b) The patentee's right is exhausted when he or she sells variety material containing a patented gene. Any subsequent use of that material or anything derived from it (progeny) is patent-free.
   c) The patentee's right is exhausted by sale of variety material containing a patented gene in respect of that material and its immediate (first-generation) progeny. Seed is sold with the implied right to reproduce it once. However, further reproduction amounts to unlicensed manufacture in competition with the patentee, and is not permitted.
   d) The farmers' exemption, as it is foreseen in the variety protection legislation, applies to all (or specified) plant varieties regardless of whether any patents relate to the variety or parts thereof. This approach is used in the European Union's Directive on the Legal Protection of Biotechnological Inventions.
   e) The patentee retains rights to control reproduction of a patented gene until he or she authorizes its use in a commercial plant variety. From then on, the plant breeder's exemption has precedence, so that anyone can take the commercial plant variety and use it in breeding.
   f) The patentee retains the right to control the use of the patented gene in further commercial breeding, but his or her right is exhausted when material containing the gene is sold to farmers. Farmers benefit from a

general 'farmer's privilege', and may not be sued for reproduction or sale of material containing patented genes where this does not amount to commercial breeding.

While other interface conditions can be envisaged,[36] these seem to be the main ones. Options (b), (e) and (f) will likely be challenged as not meeting the TRIPs requirements, in particular Article 30, because they do not give the patentee of the gene the customary rights that a patent bestows.

However, given that plants as a whole, including plant varieties, may be excluded from patentability, one might argue (1) that parts of plants may be excluded from patentability too, and (2) that the right to exclude plants from patentability includes the right to refuse patent claims that would finally extend to plants — the right to give plants immunity under patents.

3. PIC and the breeders' exemption

Access to plant varieties, in particular to local and traditional varieties, may depend on authorization from communities or nation-states, e.g., by a procedure involving PIC on which mechanisms for benefit-sharing can be built. Under plant variety protection, on the other hand, breeders are generally allowed to use plant varieties as an initial source for further breeding under the breeders' exemption. Thus, there may arise the problem that acts allowable under plant variety protection will be restricted by access legislation that may include limits to further use by third parties.

One may argue that questions related to access should be kept separate from plant variety protection. This means that the breeders' exemption can remain fully valid under plant variety protection. The question whether breeders would need PIC under national access legislation in order to use the protected variety, or before commercializing a new variety derived from the protected variety, would then be treated exclusively under access legislation. Such an approach would suggest not to deal with PIC in plant variety legislation at all or, if so, only to create a link between the two, for example, by making PIC a protection requirement for plant variety protection.

Another solution would be to exempt certain species (or varieties, uses) from the need for PIC.

A separation line could be set up according to species list (comparable to UPOV 1978), or a variety list. Species (varieties) are either on the plant variety protection list and then available for further breeding without restrictions, or they are not on this list and then require normal access and PIC procedures. Such a solution may match well with a MUSE system; it may be based on an identical, internationally agreed list.

It seems also possible to make an exemption from the PIC procedure when a certain use (e.g., plant breeding) is intended. However, this may be very difficult to control in practice.

Where the two principles are not separated, i.e., where there is a need for authorization of further use according to access legislation, and at the same time a plant variety protection system with a clear breeders' exemption, this exemption could at least partially be saved by making the need to obtain authorization a time-limited exception to the breeders' exemption.

While the solution discussed under 2.A (separation according to species) might fit best with an open multilateral system for exchange (MUSE) of plant genetic resources for food and agriculture, time-limited exemptions from the breeders' privilege as discussed under Part Three may also be used to protect 'community varieties' from being incorporated into commercial breeding programmes without consent of the community until a certain period of time has elapsed.

Abbreviations

ACTS	African Centre for Technology Studies
CBD	Convention on Biological Diversity
CCD	Convention to Combat Desertification
CGIAR	Consultative Group on International Agricultural Research
CGRFA	Commission on Genetic Resources for Food and Agriculture
COP-CBD	Fourth Conference of the Parties to the CBD
DHF	Dag Hammarskjöld Foundation
ECOSOC	Economic and Social Council of the United Nations
FAO	Food and Agriculture Organization of the United Nations
FAO-COC	Food and Agriculture Organization of the United Nations' Code of Conduct for Plant Germplasm Collecting and Transfer
GATT	General Agreement on Tariffs and Trade
GBS	Global Bio-Collecting Society
ICESR	International Convention on Social and Cultural Rights
IDRC	International Development Research Centre
ILO 169	International Labour Organisation's Convention concerning Indigenous and Tribal Peoples in Independent Countries
IPGRI	International Plant Genetic Resources Institute
IU on PGRFA	International Undertaking on Plant Genetic Resources for Food and Agriculture
OAU	Organization of African Unity
PIC	prior informed consent
PGRFA	plant genetic resources for food and agriculture
PVP	plant variety protection
TRIPs	The Uruguay Round Agreement on Trade-Related Intellectual Property Rights
TWN	Third World Network
UNCED	UN Conference on Environment and Development
UPOV	International Union for the Protection of New Varieties of Plants
VCU	value in cultivation and use
WIPO	World Intellectual Property Organization
WTO	World Trade Organization

Notes

Topic 1, Section 1, pp 3–4

1. Plant genetic resources for food and agriculture (PGRFA) have been extensively exchanged for millennia, creating a situation where all regions and countries of the world are highly interdependent for access to the genetic diversity that ensures the continued viability of major food crops. In recognition of this free flow of genetic material that is the basis of global food security, a multilateral system (MLS) for access and benefit sharing is being established as the cornerstone of the IU on PGRFA; an international agreement is currently being negotiated in FAO by its Commission on Genetic Resources for Food and Agriculture (CGRFA).
2. See Topic 3, Section 2, Articles 12 and 13, and Topic 3, Section 3, Article 19.
3. See Appendix, 'An introduction to access laws and their relationship to intellectual property laws'.

Topic 1, Section 2, pp 7–32

4. See, for example, the European Cooperative Programme on Crop Genetic Resources Networks (ECP/GR), West Asia and North Africa Plant Genetic Resources Network (WAWANET), Southern African Development Community Plant Genetic Resources Centre (SPGRC), Regional Committee for Southeast Asia (RECSEA), Australian and New Zealand Network of Plant Genetic Resource Centres (ANZNPGRC), Amazonian Network of Plant Genetic Resources (TROPIGEN), Andean Network of Plant Genetic Resources (REDARFIT), Central American Network of Plant Genetic Resources (REMERFI), Programa Cooperativo para el Desarrollo Tecnologico Agropecuario del Cono Sur (PROCISUR).
5. See *Seeding Solutions*, Vol. 1, p 8 for a general discussion of the reduction in different kinds of biological diversity.
6. Compare Articles 11–14 of Composite Draft Text of the International Undertaking on Plant Genetic Resources incorporating the Chairman's Elements (CGRFA/CG-4/00/2).
7. Executive Order No. 247 dated 18 May 1995, 'Prescribing guidelines and establishing a regulatory framework for the prospecting of biological and genetic resources, their by-products and derivatives, for scientific and commercial purposes, and for other purposes'.

Topic 2, Section 1, pp 35–41

8. See *Seeding Solutions*, Vol. 1, pp 9–10.

Topic 2, Section 2, pp 45–57

9. See *Seeding Solutions*, Vol. 1, p 8 for a general discussion of the reduction in biological diversity.

10. See the discussion in *Seeding Solutions*, Vol. 1, pp 9–10 regarding the indicators of the accelerating rate of disappearance of indigenous cultures, languages and biodiversity.
11. Many countries' laws categorize intellectual property as a kind of personal property. It is somewhat inaccurate, therefore, to put intellectual property and personal property in separate headings, as they are entirely different things.
12. See, for example, the 'Research principles for community-controlled research with the Inuit Tapirisat of Canada', developed by the Inuit Taparisat of Canada, and the 'Guidelines for conduct of participatory community research to document traditional ecological knowledge for the purpose of environmental assessment and environmental management', developed by the Dene Cultural Institute [Grenier, L. 1998. *Working with indigenous knowledge: a guide for researchers*. IDRC, Ottawa. Appendix 1, pp 87–97].
13. 'Prior art' is a term used in patent law to describe inventions or innovations that already exist and that are disclosed in legally prescribed ways to the public. In intellectual property laws that include novelty as a condition of protection, the existence of prior art means that no one can get intellectual property protection pursuant to that law for something identical with that prior art. Novelty is used in plant variety protection and patent laws. See Topic 3, Section 2, Article 8 and Topic 3, Section 3, Article 16 for details about the significance of these terms in the context of those laws. It might also be included as a condition for protection in a *sui generis* intellectual property law for the protection of indigenous and local knowledge. See Topic 2, Section 3, Article 10 for a discussion about how novelty might (or possibly should not) be included in such a law.
14. Drahos, P. 2000. 'Indigenous knowledge, intellectual property and biopiracy: is a global bio-collecting society the answer?' EIPR. pp 245–9.
15. *Seeding Solutions*, Vol. 1, p 85.

Topic 2, Section 3, pp 61–124

16. *New Shorter Oxford English Dictionary* [CD-ROM]. 1997. Oxford University Press, Oxford, UK.
17. See Topic 3, Section 2, Articles 12 and 13, and Topic 3, Section 3, Article 19.
18. Article 1(b) of ILO 169 states: "Indigenous describes those individuals and populations that have descended from populations that inhabited a country or area within a country at the time of conquest, or colonization, or the establishment of present state boundaries and who, irrespective of their legal status, retain some or all of their own social, economic, cultural and political institutions."

    The only international treaty that provides any guidance as to what might be meant by 'local' is the ILO 169. It provides the following definition of 'tribal', which may or may not contain elements that a national government would wish to include in a definition of 'local' in its own laws. Article 1(a) of the ILO 169 defines tribal peoples as those "whose social, cultural and economic conditions distinguish them from other sections of the national community, and whose status is regulated wholly or partially by their own customs or traditions or by special laws or regulations." Article 1(2) states: "Self-identification as indigenous or tribal shall be regarded as a fundamental criterion for determining the groups to which the provisions of the Convention apply."

Paragraphs 379 and 380 of the Cobo study state that:

> indigenous peoples consider themselves distinct from other sectors of the societies now prevailing in those territories. They form at present non-dominant sectors of society and are determined to preserve, develop, and transmit to future generations their ancestral territories and their ethnic identity, as the basis of their continued existence as peoples, in accordance with their own cultural matters, social institutions and legal systems. This historical continuity may consist of the continuation for an extended period reaching into the present, or one or more of the following factors:
>
> a) occupation of ancestral lands, or part of them;
> b) common ancestry with the original occupants of these lands;
> c) culture;
> d) language;
> e) residence in certain parts of the country or world;
> f) other relevant factors.

The principle of self-identification is also included in the Cobo report.

Article 1 of the proposed Inter-American Declaration on the Rights of Indigenous Peoples states:

1. In this Declaration indigenous peoples are those who embody historical continuity with societies which existed before the conquest and settlement of their territories by Europeans. (Alternative 1) [, as well as peoples brought involuntarily to the New World who freed themselves and re-established the cultures from which they have been torn]. (Alternative 2) [, as well as tribal peoples whose social, cultural and economic conditions distinguish them from other sections of the national community and whose status is regulated wholly or partially by their own customs or traditions or by special laws or regulations].
2. Self-identification as indigenous or tribal shall be regarded as fundamental criterion for determining the groups to which the provisions of this Declaration apply ...

The Draft Declaration on the Rights of Indigenous Peoples deliberately does not include a definition of 'indigenous'. Indigenous delegates to the negotiations of the Draft Declaration have fought hard against the introduction of such a definition, fearing that it would be used to exclude peoples who should be recognized as indigenous in different countries and restrict their right to self-determination [Tauli-Corpuz, V. 1998 Dec. 'Indigenous peoples' lobbying and advocacy in the international arena'. *Indigenous Perspectives* 1(1):30–35]. But indigenous self-identification is included as a key component of determining who is 'indigenous' pursuant to the Draft Declaration. Article 8 of the Draft Declaration states: "Indigenous peoples have the collective and individual right to maintain and develop their distinct identities and characteristics, including the right to identify themselves as indigenous and be recognized as such."

19. For traditional knowledge, there is generally no 'date of invention', since it typically arises incrementally. This makes it meaningless to discuss what would be obvious at a particular date.
20. *Foster v. Mountford & Rigby*, 14 ALR 71 (1976).
21. Canadian Patent Act, Section 2. (Identical with the US Patent Act definition of 'invention'.)

Topic 3, Section 2, pp 135–181

22. For the purposes of this calculation, we are starting from the US Plant Patent Act (1930), which was designed primarily to create intellectual property protection for flowering ornamental plants.
23. What is 'human' DNA? Is it only DNA obtained from a human being by cloning and multiplication, or does it include DNA with the same sequence of bases, but obtained from other sources, or synthesized *in vitro*? The latter will probably be physically identical with the former. Short DNA sequences in an organism are like words in a book, in that very few will be unique to the organism (or book). In fact, organisms are more like one another than most books, because organisms evolve from earlier organisms, like successive editions of a book. A long DNA sequence may be found in a range of organisms, or it may be characteristic of (only found in) a particular organism, as an extended quotation may be unique to a particular book. Even if characteristic, it may not be important. An organism (and especially the human organism) is more than the sum of its parts.
24. UPOV TG-ROM 2000 Test Guidelines at <http://www.upov.int/tg-rom/start.htm>.
25. One exception is the old patent law of the USSR — a dismal failure.
26. The rationale for protecting Farmers' Rights is set out in FAO Resolution 5/89 as "rights arising from the past, present and future contributions of farmers in conserving, improving, and making available plant genetic resources, particularly those in the centres of origin/diversity." Resolution 3/91 further states that "Farmers' Rights will be implemented through international funding on plant genetic resources, which will support plant genetic conservation and utilization programmes, particularly, but not exclusively, in the developing countries". This was adopted by the FAO Conference (175 countries) in 1991.

    Article 15 of the IU on PGRFA, currently under negotiation, defines Farmers' Rights as follows:

    15.2 The Parties agree that the responsibility for realizing Farmers'

Rights, as they relate to Plant Genetic Resources for Food and Agriculture, rests with national governments. In accordance with their needs and priorities, each Party should, as appropriate, and subject to its national legislation, take measures to protect and promote Farmers' Rights, including:

a) protection of traditional knowledge relevant to plant genetic resources for food and agriculture;
b) the right to equitably participate in sharing benefits arising from the utilization of plant genetic resources for food and agriculture;
c) the right to participate in making decisions, at the national level, on matters related to the conservation and sustainable use of plant genetic resources for food and agriculture.

15.3 Nothing in this Article shall be interpreted to limit any rights that farmers have to save, use, exchange and sell farm-saved seed / propagating material, subject to national law and as appropriate.

27. 513 US 179 (1995).
28. Rappert, B. 1995. The US extension of plant variety protection: a critical evaluation. *Science and Public Policy* 22(2):95–105.

Topic 3, Section 3, pp 185–226

29. What are 'indigenous and local communities'? What distinguishes their traditional lifestyles as particularly relevant for the conservation and sustainable use of biological diversity? A range of answers to these questions is discussed in Topic 2, Section 3 of this volume, especially in Part Two.
30. 51 Cal. 3d 120, 793 P.2d 479, 271 Cal. Rptr. 146 (9 July 1990).
31. Decision G01/98.
32. *Windsurfing International Inc. v. Trilantic Corp* (1985), 7 CIPR 281 (Fed CA).
33. *Visx Inc. v. Nidek Co. Ltd* (1995), 68 CPR (3d) 272 (Fed Ct).

Appendix, pp 227–236

34. Unless that was the practice in a state before 31 Oct 1979 and that state wishes to continue its practice after becoming a member of UPOV.
35. However, this will not help claims to DNA sequences that do *not* occur in nature: for example, to a specified plant gene under control of a specified viral promoter. Finding allowable forms of claim (perhaps a little narrower than those currently granted) might not be too serious a challenge to the ingenuity of patent attorneys.
36. For example, it has been suggested that the right to control use of the patented gene in breeding might expire when the gene has been inserted in, say, five varieties. But this discourages the patentee from putting as many diverse varieties on the market as possible.